WELL AP®
EXAM PREPARATION GUIDE

based on the WELL Building Standard v1, with Q1 2018 Addenda

AMERICAN TECHNICAL PUBLISHERS
Orland Park, Illinois

Charles A. Vescoso Jr., WELL AP

WELL AP® Exam Preparation Guide contains procedures commonly practiced in industry and the trade. Specific procedures vary with each task and must be performed by a qualified person. For maximum safety, always refer to specific manufacturer recommendations, insurance regulations, specific job site and plant procedures, applicable federal, state, and local regulations, and any authority having jurisdiction. The material contained herein is intended to be an educational resource for the user. American Technical Publishers, Inc. assumes no responsibility or liability in connection with this material or its use by any individual or organization.

American Technical Publishers, Inc., Editorial Staff

Editor in Chief:
Jonathan F. Gosse
Vice President—Editorial:
Peter A. Zurlis
Assistant Production Manager:
Nicole D. Bigos
Technical Editor:
Dane K. Hamann
Supervising Copy Editor:
Catherine A. Mini
Copy Editor:
James R. Hein
Talia J. Lambarki
Catherine A. Mini

Cover Design:
Steven E. Gibbs
Art Supervisor:
Sarah E. Kaducak
Illustration/Layout:
Bethany J. Fisher
Nick G. Doornbos
Steven E. Gibbs
Digital Media Coordinator:
Adam T. Schuldt
Digital Resources:
Cory S. Butler
Lauren M. Lenoir

3 4 5 6 7 8 9 – 18 – 9 8 7 6 5 4 3 2 1

Printed in the United States of America

ISBN 978-0-8269-1323-4

MIX
Paper from responsible sources
FSC® C132124

Printed with inks containing soy and/or vegetable oils

made in USA

ACKNOWLEDGMENTS

The authors and publisher are grateful for the photographs, technical information, and assistance provided by the following organizations:

Autodesk, Inc.
Canada Beef Inc.
Center for the Built Environment
Concept2, Inc.
CuVerro® (Olin Brass)
Delos Living, LLC
Denmarsh Photography, Inc.
Extech Instruments
Festool USA
Fresh-Aire UV
Hach Company
International Living Future Institute
International WELL Building Institute (IWBI)
Knoll, Inc.
LEDtronics, Inc.
Legend Valve and Fitting, Inc.
Linden Group Architects
Messermeister
National Garden Bureau Inc.
NREL
Phipps Conservatory and Botanical Gardens
Precor
Steelcase
StepJockey Smart Signs
Sullivan University
USDA NRCS
U.S. Green Building Council (USGBC)

Technical Reviewers

Sonali Bhasin
WELL AP
Program Development Specialist, WELL Assessor
Green Business Certification Inc.

Barbara Fanning
WELL AP, LEED Green Associate
Founder/CEO
Mindswing Consulting

Deepak Gulati
LEED AP BD+C, WELL AP
GBCI Certification Specialist, WELL Assessor
Green Business Certification Inc.

Kay Kane
LEED Green Associate
Instructional Design Specialist
U.S. Green Building Council

Jeremy R. Poling
P.E., LEED AP O+M, LEED AP BD+C, WELL AP
Senior Certification Reviewer, WELL Assessor
Green Business Certification Inc.

Megan Sparks
LEED AP O+M, BD+C, WELL AP, EIT
Director, Integration Strategy, WELL Assessor
Green Business Certification Inc.

CONTENTS

CONTENTS

CONTENTS

CONTENTS

INTRODUCTION

WELL AP® Exam Preparation Guide is a comprehensive study reference used to prepare for the WELL AP exam. This exam preparation guide provides a detailed and efficient approach to studying through the use of concise text and detailed, full-color illustrations and photos. This guide and its online learner resources are designed to complement the WELL Building Standard v1, with Q1 2018 addenda, and the *WELL Certification Guidebook* (v1, Q1 2018).

WELL AP® Exam Preparation Guide emphasizes the mastery of the key topics of the WELL Concepts and certification process and aids in exam success through the following:

- the process of registering for, studying for, and taking the WELL AP exam
- the structure, project types, and intent of the WELL Building Standard
- the relationship and value of the WELL Building Standard to human health and wellness
- the process for WELL certification, the importance of documentation and performance verification, and the roles of the project team members and the WELL assessor
- an overview of the seven WELL Concepts and the 105 features, including the requirements of each part of the features
- chapter review questions that assess the knowledge in each chapter
- WELL AP exam practice questions at the end of each chapter that help prepare for the exam
- four 100-question practice exams (one in the guide and three provided online) that prepare the candidate for the exam
- an appendix that contains an answer key and other helpful reference materials
- a glossary that contains definitions for the key terms on the WELL AP exam

To aid in the successful passing of the WELL AP exam, each element in this exam preparation guide and the digital learner resources is designed to promote quick comprehension. These elements include objectives, key terms and definitions, Wellness Facts, By the Numbers, chapter review questions, and practice exam questions.

By the Numbers present facts and statistics from the WELL Building Standard that may be encountered on the exam.

WELL AP Exam Practice Questions provide sample questions that reflect the format of the WELL AP exam.

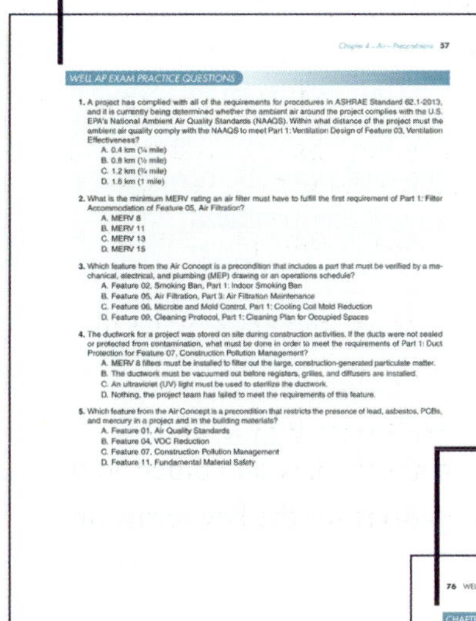

Chapter Reviews provide questions that extensively review the information in a chapter.

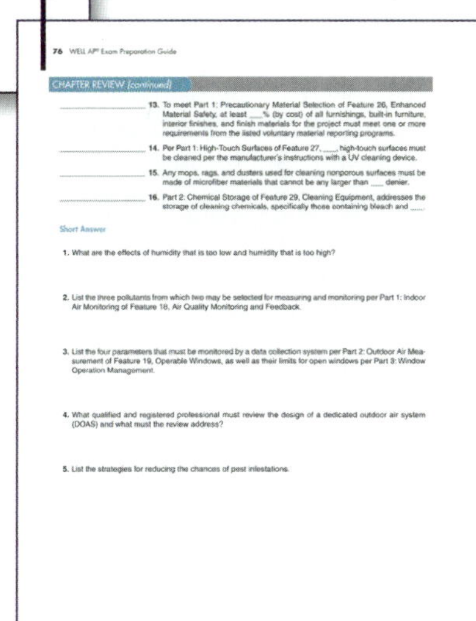

Key Terms and Definitions provide a list of terms and definitions to understand for taking the WELL AP exam.

Objectives address the main wellness principles and knowledge domains within each chapter.

Wellness Facts reinforce the impacts of the built environment on human body systems, health, and wellness.

HOW TO USE THE WELL AP® EXAM PREPARATION GUIDE

WELL AP® Exam Preparation Guide provides a comprehensive review of the wellness principles and content areas of the WELL Building Standard®. This exam preparation guide is based on the Q1 2018 versions of the WELL Building Standard v1 and *WELL Certification Guidebook* and reflects the knowledge and content areas that will be tested by the WELL AP exam. Exam candidates should check the WELL website (www.wellcertified.com) from the International WELL Building Institute (IWBI) for information regarding updates to the WELL AP exam.

By using and studying *WELL AP® Exam Preparation Guide* in conjunction with the primary references and exam specifications, an exam candidate can achieve exam day success. This exam preparation guide is divided into the following chapters and sections:

- Chapter 1 explains the process of registering for, studying for, and taking the WELL AP exam.
- Chapter 2 introduces the WELL Building Standard, explains its structure, identifies the Concepts within it, explains various synergies and tradeoffs, describes the project types and pilot standards, and explains the principles of wellness and human health in the built environment.
- Chapter 3 explains the value of WELL certification, the process for projects seeking certification in the WELL Building Standard, the importance of documentation and performance verification, and how the different levels of WELL Certification are achieved.
- Chapters 4 through 11 provide a comprehensive overview of each Concept contained within the WELL Building Standard and the principles of wellness that shape the strategies for meeting the requirements of the Features within each Concept.
- Chapter 12 is a 100-question practice exam.
- The Chapter Review and WELL AP Exam Practice Questions at the end of each chapter in this exam preparation guide allow an exam candidate to assess knowledge learned and determine areas where further study is needed.
- The Appendix provides an Answer Key with a comprehensive list of answers to all of the questions in this guide as well as other helpful reference materials.
- The Glossary contains definitions for all of the key terms used in this guide that may be on the WELL AP exam.

The online learner resources that accompany the *WELL AP® Exam Preparation Guide* provide further assessment tools, such as three additional 100-question practice exams, Quick Quizzes®, a By the Numbers quiz, and flash cards, all of which can be used to achieve exam-day success. Directions for accessing the online learner resources are found on the Learner Resources page directly preceding Chapter One.

UNDERSTANDING THE KNOWLEDGE DOMAINS

The WELL AP exam contains 100 multiple-choice questions, which includes 15 pretest questions, that reflect the knowledge domains listed in the *WELL AP Candidate Handbook*. For the benefit of the exam candidate and to enhance the educational aspects of this exam preparation guide, these domains and where they are addressed are detailed below.

The knowledge domains for the WELL AP exam reflect what an exam candidate must know about the WELL certification process and the Concepts that compose the WELL Building Standard. Exam candidates should consult the *WELL AP Candidate Handbook* for the complete list of knowledge domains and the details of their component. The knowledge domains, the number of exam questions each domain represents, the domain covered by each of the specific chapters of this exam preparation guide, and the general components of each domain include the following:

- **Air (13 questions)** *Covered in Chapters 4 and 5*
  - Human health
  - Strategies
  - Operations
- **Water (8 questions)** *Chapter 6*
  - Human health
  - Treatment and management
- **Nourishment (8 questions)** *Chapter 7*
  - Human health
  - Strategies
  - Design, operations, and management
- **Light (11 questions)** *Chapter 8*
  - Human health
  - Metrics and technical
  - Strategies
- **Fitness (8 questions)** *Chapter 9*
  - Human health
  - Strategies
- **Comfort (10 questions)** *Chapter 10*
  - Acoustic
  - Thermal
  - Ergonomics
- **Mind (9 questions)** *Chapter 11*
  - Human health
  - Stress reduction
  - Transparency
  - Beauty and biophilia
  - Adaptable spaces: design and policy
- **WELL Certification (10 questions)** *Chapters 2 and 3*
  - Planning and preparation
  - Execution for WELL Certification
  - Advocacy and promotion of WELL
- **Synergies (8 questions)** *Chapter 2*
  - Conflicts and tradeoffs
  - Application and education

LEARNER RESOURCES

WELL AP® Exam Preparation Guide includes access to online learner resources that reinforce guide content and enhance learning. These online resources can be accessed using either of the following methods:

- Key ATPeResources.com/QuickLinks into a web browser and enter QuickLinks™ Access code **935534**.
- Use a Quick Response (QR) reader app to scan the QR Code with a mobile device.

Learner Resources
ATPeResources.com/QuickLinks
Access Code: 935534

The online learner resources include the following:

- **Quick Quizzes®** that provide interactive questions for each chapter, with embedded links to highlighted content within the guide and to the Illustrated Glossary
- **Illustrated Glossary** that serves as a helpful reference to commonly used terms, with selected terms linked to illustrations in the guide
- **Flash Cards** that provide a self-study/review tool for exam preparation
- **Practice Exams** that provide opportunities for knowledge-retention assessment and exam-taking preparation
- **By the Numbers Quiz** that provides interactive questions for additional facts and statistics from the guide
- **Media Library** that consists of videos and animations that reinforce content in the guide
- **ATPeResources.com**, which provides access to additional online resources that support continued learning

To obtain more information on other related exam preparation material, including the ATP*Web*Book™ for this title, visit the American Technical Publishers website at www.atplearning.com.

The Publisher

CHAPTER ONE

THE WELL AP EXAM

The WELL AP exam tests an exam candidate's knowledge of and experience with the WELL Building Standard. Before attempting to take the exam, candidates should ensure that they are familiar with the requirements and references for the exam. A good study plan can also be used to ensure success on exam day.

KEY TERMS

- Green Business Certification Inc. (GBCI)
- WELL Accredited Professional (AP)
- WELL Online

OBJECTIVES

- Identify the eligibility requirements for taking the WELL AP exam.
- Describe the format of the WELL AP exam.
- Explain how the knowledge domains apply to the WELL AP exam.
- Identify the references used for the WELL AP exam.
- Identify the process for studying for the WELL AP exam.
- Identify the process for taking the WELL AP exam.
- List the requirements for maintaining the WELL AP credential.
- Identify the four activities that qualify for continuing education (CE) hours.

Learner Resources
ATPeResources.com/QuickLinks
Access Code: 935534

BECOMING A WELL AP

As more businesses and organizations seek WELL Building Standard certification for their buildings, the need for qualified professionals who can actively lead these building projects increases. A *WELL Accredited Professional (AP)* is an individual who possesses the knowledge and skill necessary to support the WELL certification process. The WELL AP credential validates not only a person's competency with the WELL Building Standard but also that person's knowledge in human health and wellness in the built environment.

Professionals who have earned their WELL AP credentials are valuable to employers because they help businesses become more competitive, improve communication through shared understanding of the WELL Building Standard, and show commitment to wellness and sustainability. Professionals in architecture, construction, building maintenance, engineering, real estate, commissioning, product development and manufacturing, and healthcare can all benefit from earning the WELL AP credential. Because the WELL AP exam tests an exam candidate's understanding of the WELL Building Standard and the WELL certification process, professionals who have passed the WELL AP exam demonstrate competency in their ability to perform the duties of a WELL AP.

THE WELL AP EXAM

The content of the WELL AP exam was developed by leading subject matter experts in the fields of building design and health and wellness. Because the exam is computer-based, it will undergo periodic validation by the subject matter experts and be reviewed and updated on an annual basis. *Note:* The current iteration of the WELL AP exam is based on the WELL Building Standard v1 with Q1 2018 addenda.

The WELL AP exam and credential are administered by the Green Business Certification Inc. *Green Business Certification Inc. (GBCI)* is a third-party organization that provides independent oversight of professional credentialing and project certification programs related to green building and health and wellness in the built environment. GBCI also provides WELL APs with policies and procedures for maintaining their credentials through the Credential Maintenance Program (CMP).

Registering for the Exam

To register for the WELL AP exam, a person must first create an account on WELL Online. *WELL Online* is the web-based portal for registering for the WELL AP exam and for completing the WELL certification process. A person can then sign in to WELL Online and click "WELL AP" under the Account tab. This will lead to a page with a button to register for the exam. It is important that the name entered for registration appears exactly as it does on the ID that the exam candidate will provide for identification at the test center. *Note:* The registration application requires that the Latin alphabet (such as English) be used for a candidate's name and ID.

After registering for the exam, a candidate will be redirected to the Prometric website. Prometric is a third-party organization that proctors the WELL AP exam as well as many other credentialing exams. There, the exam candidate can schedule the exam date and Prometric test center location. *Note:* Upon scheduling the date and location, an exam candidate will be emailed a confirmation number that must be recorded in case the candidate must communicate with Prometric.

Eligibility Requirements. Exam candidates must be 18 years of age or older in order to take the WELL AP exam. Candidates must also agree to the Disciplinary and Exam Appeals Policy provided by GBCI. No previous experience with WELL or health and wellness in the built environment is required. However, exposure to those concepts and principles is strongly recommended by GBCI.

A valid, unexpired, official ID must be presented at the Prometric test center on the day of the exam. The ID must have a photo that looks like the exam candidate and a signature. Acceptable IDs include, but are not limited to, driver's licenses, passports, military IDs, and alien IDs/resident alien cards. If the ID has a photo but no signature, the exam candidate can also present a credit card with a signature in the same name.

Exam Fees. The fees for the WELL AP exam are an investment in an exam candidate's future in the industry of health and wellness in the built environment. The cost of the WELL AP exam varies depending on whether the exam candidate or the employer of the candidate is a member of the U.S. Green Building Council (USGBC) or the American Society of Interior Designers (ASID), whether the exam candidate currently holds a LEED AP or LEED Green Associate credential, and the exam location. Up-to-date costs can be found on the WELL website.

The WELL AP exam can be rescheduled or canceled through Prometric for a full refund up to 30 days before the scheduled exam day. If there are less than 30 days but more than 3 days before the scheduled exam day, an exam candidate will receive a full refund minus a $50 fee. An exam candidate will not be allowed to reschedule an exam if there are 2 days or less to the scheduled exam day, and no refund will be granted.

Exam Format

The WELL AP exam consists of 100 multiple-choice questions, including 15 pretest questions. The exam is scored between 125 and 200, with a passing score of 170 or higher. The exam must be completed in a 2 hour period. There is also an optional 10 minute tutorial and an optional 10 minute exit survey. An exam candidate's score is immediately available following completion of the exam. Once an exam candidate receives a passing score, the certificate for the WELL AP credential will be available for download on WELL Online.

Exam candidates should endeavor to answer all questions on the exam, since the exam includes both scored and unscored questions. The scored and unscored questions are unmarked and randomly delivered. The unscored questions are used to gather data and determine whether those questions will appear on future versions of the exam. Exam candidates may also choose to leave comments regarding exam questions by using the comment button on the screen of their computer station.

Since the WELL AP exam is a computer-based exam, an exam candidate will view and answer the questions on the computer screen. The computer records the answers and keeps track of the 2 hour time limit. An exam candidate can also navigate between questions using the computer. This allows the exam candidate to flag and return to unanswered questions or to change answers.

Types of Questions. There are three types of questions that test an exam candidate's critical and analytical thinking capabilities. The three types of questions are recall questions, application questions, and analysis questions. All 100 multiple-choice questions on the exam fall into one of these types. Recall questions assess whether a candidate can remember factual information derived from the primary references of the exam. Application questions assess whether a candidate can solve a unique problem using familiar solutions from the primary references. Analysis questions assess whether a candidate can evaluate and break down a problem to create a solution based on the knowledge learned from the primary references.

Exam Specifications

Exam candidates must be familiar with the exam specifications that are provided in the *WELL AP Candidate Handbook*. **See Figure 1-1.** The exam specifications include a general description of the content areas of the WELL AP exam, which are called knowledge domains. This section of the candidate handbook also specifies how many questions for each content area will be found on the exam. A list of references that an exam candidate can use as study materials is at the end of the exam specifications.

WELL AP CANDIDATE HANDBOOK

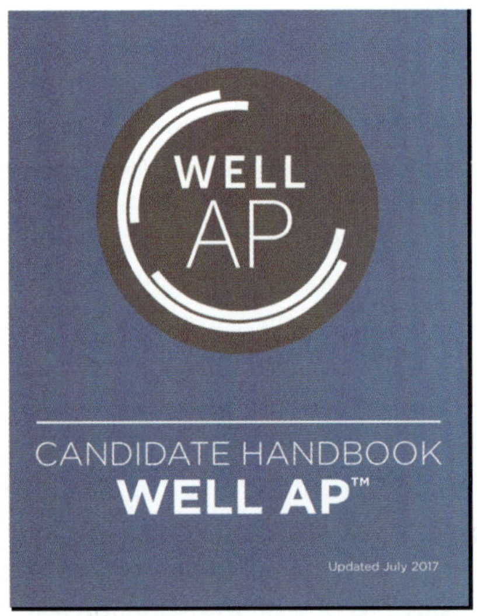

Figure 1-1. The *WELL AP Candidate Handbook* includes the exam specifications for the WELL AP exam and should be reviewed thoroughly by an exam candidate.

Knowledge Domains. The knowledge domains reflect the knowledge that an exam candidate must understand about the WELL Building Standard and the WELL certification process. The knowledge domains include nine sections that are further itemized into detailed content areas. Exam candidates should consult the *WELL AP Candidate Handbook* for the specific details of each knowledge domain. The following is a breakdown of the number of questions about each knowledge domain represented on the exam:

- Air Concept (13 questions)
- Water Concept (8 questions)
- Nourishment Concept (8 questions)
- Light Concept (11 questions)
- Fitness Concept (8 questions)
- Comfort Concept (10 questions)
- Mind Concept (9 questions)
- WELL certification (10 questions)
- Synergies (8 questions)

Primary References. Exam candidates must be familiar with the references listed in the *WELL AP Candidate Handbook*. **See Figure 1-2.** These references serve as a good foundation for studying for the WELL AP exam. Besides the candidate handbook itself, exam candidates should study the following primary references:

- **WELL Building Standard v1 with Q1 2018 Addenda.** The exam is primarily based on the WELL Building Standard from the International WELL Building Institute (IWBI). Currently, exam content covers up to the Q1 2018 addenda. Exam candidates should consult the WELL website for updates or changes to exam content.
- **WELL Certification Guidebook (v1, Q1 2018).** The *WELL Certification Guidebook* provides an in-depth explanation of the process and practices for WELL certification. Exam candidates should consult the WELL website for updates or changes to exam content.

Studying for the WELL AP Exam

The amount of time needed to study, and the study process, depends on the background, education, and experience of the individual exam candidate. A candidate with extensive knowledge and experience working with WELL projects and the principles behind the WELL Building Standard may require less intensive study than someone who is not as familiar with the specifics of WELL.

Figure 1-2. The primary references for the WELL AP exam include the WELL Building Standard v1 and the *WELL Certification Guidebook*.

However, it is important for exam candidates to study the primary references regardless of their experience with the WELL Building Standard. The primary references introduce specific themes and language that the exam candidates must know to pass the exam. An individual with no familiarity with the WELL Building Standard and little background in construction, design, or human health should expect to spend about three months studying for the exam if that individual works or is a full-time student. **See Figure 1-3.**

The first thing for an exam candidate to do is register for the WELL AP Exam. By registering and setting an exam date, a study plan can be developed. A three-month study plan can be shortened or lengthened depending on the exam candidate. An end-to-end study plan is also available on the WELL website (www.wellcertified.com). Also, the date of the exam may be changed later if needed.

Study Plan Methodology. The WELL AP exam mainly requires exam candidates to recall the organization and details concerning the WELL Building Standard and the WELL certification process. Once candidates understand these exam materials, they can begin other studying strategies, such as memorizing important WELL facts and figures. Memorization can be an effective strategy for exam day success. While memorization is an important part of studying for the exam, candidates must strive to understand the principles, strategies, and methodology of the WELL Building Standard. Another key is to learn how the features in the WELL Concepts interact with or affect one another, and the strategies that can be used to meet the requirements of the features.

WELL AP EXAM SAMPLE STUDY PLAN

MONTH 1	MONTH 2	MONTH 3
• READ *WELL AP CANDIDATE HANDBOOK*	• CONTINUE READING *WELL AP EXAM PREPARATION GUIDE*	• CONTINUE READING *WELL AP EXAM PREPARATION GUIDE*
• REGISTER FOR WELL AP EXAM	• COMPLETE CHAPTER REVIEW QUESTIONS AND PRACTICE EXAM QUESTIONS FOR EACH CHAPTER	• COMPLETE CHAPTER REVIEW QUESTIONS AND PRACTICE EXAM QUESTIONS FOR EACH CHAPTER
• BEGIN READING PRIMARY REFERENCES		• COMPLETE PRACTICE EXAM QUESTIONS IN CHAPTER 12
• BEGIN READING *WELL AP EXAM PREPARATION GUIDE*		• REVIEW ALL MATERIALS
		• TAKE AND PASS WELL AP EXAM

Figure 1-3. A solid study plan based on the experience and knowledge of the exam candidate can increase the chances of successfully passing the WELL AP Exam. The WELL website (www.wellcertified.com) also provides an end-to-end study plan.

Study Materials. In addition to studying the primary references, exam candidates can use other study materials such as review questions, practice questions, practice exams, and study cards. This exam preparation guide includes chapter review questions at the ends of Chapters 2 through 11. While there are no official IWBI or GBCI practice exams, the WELL AP exam practice questions also located at the ends of Chapters 2 through 11 are written to familiarize an exam candidate with both the question material and the question format. Chapter 12 is a practice exam that consists of 100 multiple-choice questions.

Additionally, three digital 100-question practice exams and a 46-question By the Numbers exam are located in the *WELL AP Exam Preparation Guide* Learner Resources, which are accessible via a QR code reader app or web browser.

To best use this exam preparation guide, it is recommended that an exam candidate take the practice exam after reviewing each chapter and completing the chapter review questions and WELL AP exam practice questions. Any incorrect answers on the practice exam can then be used to determine which materials or chapters need to be re-reviewed.

An exam candidate may also choose to divide the practice exam into smaller sections to concentrate learning time. For example, instead of spending 2 hours to answer 100 practice questions, a candidate may choose to spend 1 hour answering 50 practice questions and then another hour finishing the exam at a later time. An exam candidate should use the practice exam and practice questions until comfortable answering a vast majority of the questions.

Study Tips. There are several study tips that can be used to make studying for the WELL AP exam easier and more productive. These tips include the following:

- having a positive attitude and confidence
- dividing the reference materials into smaller sections for easier studying and arranging a manageable daily study schedule
- setting aside time (at least 30 minute increments) that is dedicated only to studying
- finding a quiet, comfortable place to study, away from distractions
- turning off cell phones and other electronic devices
- informing others when studying to reduce interruptions

- dividing up the study time and taking small breaks to mentally recharge
- using study groups or studying with other exam candidates
- speaking to others who previously took the exam

Mnemonic or Other Memory Devices. The creation and use of mnemonic or other memory devices can help some exam candidates remember certain information pertaining to the WELL Building Standard. For example, the word CLAM can help candidates recall the four dissolved metals for which a water supply must be tested quarterly per Feature 35, Periodic Water Quality Testing: *C*opper, *L*ead, *A*rsenic, *M*ercury.

Exam candidates can also memorize the string of numbers 12-34-45-56-65-76-88 to help recall the cutoffs for preconditions and optimizations for each of the WELL Concepts: Air—Feature 12, Water—Feature 34, Nourishment—Feature 45, Light—Feature 56, Fitness—Feature 65, Comfort—Feature 76, and Mind—Feature 88. In general, when a specific concept is being discussed, a feature is a precondition if it has a number lower than one of these given numbers within the concept. If the number is higher, then the feature is an optimization. For example, in the Air Concept, Feature 10 is a precondition and Feature 26 is an optimization.

A few features can be either preconditions or optimizations depending on the project type. However, exam questions generally ask about features that are either only a precondition or an optimization regardless of the project type.

Taking the WELL AP Exam

The day of the WELL AP exam can be stressful. It is important to eat breakfast if the exam is in the morning or to have a light lunch if the exam is in the afternoon. Hunger can cause a lack of focus, which increases stress. It is also important to arrive at the test site 30 minutes to 60 minutes ahead of the scheduled exam time. This will allow time to find the location, and any extra time can be used to relax. If a candidate is traveling from out of town, it would be beneficial to visit the exam site a day or two before the exam to eliminate any difficulty in finding the location. Exam candidates must bring a valid ID that has a signature, photograph, and expiration date to the test center. The exam candidate's name on the exam application and ID must match.

A test administrator provides exam candidates with a dry-erase board with markers or a pencil and blank paper before escorting them to their workstations. Once exam candidates are sitting at their own terminal, they may create a reference by writing whatever they want on this board or paper before beginning the exam.

Although candidates have 2 hours to answer the exam questions, an additional 20 minutes is allocated to the exam session. At the beginning of the exam, there is a 10-minute tutorial on how to take the exam. At the end of the exam, there is a 10-minute survey. The 10-minute tutorial can be important for understanding how to use the testing interface. For example, the tutorial teaches how to highlight or cross out portions of the exam questions or answers. The tutorial can also teach a candidate how to flag questions for later review.

There are several tips that can be followed to relieve stress and help efficiently complete the WELL AP exam in the allotted time. These tips include the following:

- Read each question completely, and read every possible answer of the multiple-choice questions. Make sure that the question is clearly understood. Use the highlight and cross-out features to narrow the choices.
- First answer all of the questions with known answers, and flag the questions that are left unanswered. During the exam, it is possible to go back to unanswered questions anytime. Often a question is answered by a following question.

- Allow time to finish the exam. If time permits, go back and answer any flagged, unanswered questions.
- If the answer to a question is unknown, take an educated guess. Do not leave any unanswered questions since these will be marked as incorrect.
- Be cautious when changing answers. Usually the first answer is the correct one.
- Do not become discouraged if unsure of too many answers, as the results of the exam may be difficult to predict and a passing score may be achieved despite uncertainty.

At the end of the allotted time or once the exam has been finished, a candidate's score is calculated by the computer. The score will appear on the screen, and the candidate will be given a printed copy. A score of 170 is required to pass the WELL AP exam.

WELL AP CREDENTIAL MAINTENANCE

Once the WELL AP credential is earned, WELL APs must maintain their credential by earning a minimum number of continuing education (CE) hours every two years. There is also a credential maintenance fee that must be paid. Failure to complete or report CE hours for a two-year period will result in the expiration of the credential. Once a credential has expired, a former WELL AP must take and pass the WELL AP exam again to be reinstated.

WELL APs are required to have 30 CE hours, but these can be earned in a variety of ways. Of the 30 CE hours, at least 6 hours must be WELL-specific CE hours. WELL-specific CE is defined as an activity that has a connection to some aspect of the WELL Building Standard and the relevant health and wellness science and research related to WELL topics. WELL-specific content covers at least one WELL Concept and includes pertinent health research/impacts based on the latest standards, guidelines, and scientific research.

The GBCI's credential maintenance program (CMP) is designed to allow professionals the opportunity to maintain credentials for both WELL and the USGBC's Leadership in Energy and Environmental Design™ (LEED®) green building program. A WELL AP may choose to earn the other 24 CE hours by participating in activities that could be used for either LEED or WELL credential maintenance. Four activities can be used to earn CE hours. These activities include education, project participation, authorship, and volunteering. **See Figure 1-4.**

WELL AP Credential Maintenance		
Activity	**Amount of CE Hours**	**Qualifying Activities**
Education 	Unlimited	Courses and Presentations: • College and university courses • IWBI workshops, courses, and events • Greenbuild International Conference and expo • WELL-specific courses on Education@USGBC
LEED and WELL Project Participation 	Unlimited	LEED and WELL Project Participation: • LEED projects = 1 CE hour per credit • WELL projects = 1 CE hour per feature • Project administrators = 2 CE hours
Authorship 	Unlimited	• Print or digital publication article = 3 CE hours for published article • Print or digital publication article = 10 CE hours for published book • LEED and WELL specificity determined on case-by-case basis
Volunteering 	No more than 50% of total CE hours	• Participation on USGBC or GBCI board of directors, steering committees, or working groups • LEED and/or WELL credential exam development and maintenance activities • Local USGBC chapter boards and committees • Volunteer organizations that support LEED and WELL

Figure 1-4. Four activities can be used to earn the required 30 continuing education (CE) hours for WELL AP credential maintenance: education, project participation, authorship, and volunteering.

KEY TERMS AND DEFINITIONS

Green Business Certification Inc. (GBCI): A third-party organization that provides independent oversight of professional credentialing and project certification programs related to green building and health and wellness in the built environment.

WELL Accredited Professional (AP): An individual who possesses the knowledge and skill necessary to support the WELL certification process.

WELL Online: The web-based portal for registering for the WELL AP exam and for completing the WELL certification process.

CHAPTER TWO

THE WELL BUILDING STANDARD

A dynamic rating system for the built environment that combines best practices in design and construction with evidence-based health and wellness interventions, the WELL Building Standard® is a significant new development for the building industry. The WELL Building Standard is a culmination of research and development by leading professionals in medicine, science, and the green building and construction industry. The overall aim of the WELL Building Standard is to improve and protect the health and wellness of building occupants.

OBJECTIVES

- Explain the organization of the WELL Building Standard.
- Describe each of the seven concepts in the WELL Building Standard.
- Differentiate between a precondition and an optimization.
- Describe the synergies between the WELL Building Standard and other green building and sustainability programs.
- Describe the synergies and trade-offs between the WELL Concepts and features.
- Identify the project types and the number of preconditions and optimizations for each.
- Identify the pilot standard programs and to which space types they apply.
- Explain how the WELL Building Standard aims to protect the health of human body systems in the built environment.

KEY TERMS

- feature
- International WELL Building Institute (IWBI)
- optimization
- part
- precondition
- synergy
- trade-off
- WELL Building Standard

Learner Resources
ATPeResources.com/QuickLinks
Access Code: 935534

WELL

The *WELL Building Standard*, simply referred to as WELL, is a performance-based system for measuring, certifying, and monitoring features of the built environment that impact human health and well-being through air, water, nourishment, light, fitness, comfort, and mind. WELL is the first standard of its kind to focus solely on human health and wellness in regards to the built environment.

Delos Living LLC, a real-estate and technology firm that pioneers designs for wellness-centered built environments, founded the WELL Building Standard and established the International WELL Building Institute™ (IWBI™). The *International WELL Building Institute (IWBI)* is a public-benefit organization that administers the WELL Building Standard and aims to improve human health and well-being through the built environment.

WELL BUILDING STANDARD ORGANIZATION

The WELL Building Standard is organized into seven major categories of wellness called concepts. These concepts are then divided into features. A *feature* is one of the 105 sections of the WELL Building Standard with a specific health intent. Five of the features, located at the end of the WELL Building Standard, promote innovation and are not associated with a specific concept.

The seven WELL Concepts are the factors that impact the health of building occupants. WELL includes the Air Concept, Water Concept, Nourishment Concept, Light Concept, Fitness Concept, Comfort Concept, and Mind Concept, as well as the Innovation features. **See Figure 2-1.**

Air Concept

The aim of the Air Concept is to achieve optimal indoor air quality to support the health and well-being of building occupants. The features of the Air Concept promote strategies for achieving clean air through the removal of airborne contaminants, pollution prevention, and air purification. The Air Concept is the largest concept in WELL with 29 features.

Water Concept

The aim of the Water Concept is to optimize the quality of water available to building occupants and promote its accessibility. The features in the Water Concept promote strategies for ensuring that water is safe, clean, and easily accessible through filtration, treatment, and strategic placement. The Water Concept includes 8 features.

Nourishment Concept

The aim of the Nourishment Concept is to encourage healthier eating habits and food culture that lead to the better health and well-being of building occupants. The features in the Nourishment Concept promote strategies for healthier eating habits by increasing the availability and visibility of fresh foods, limiting the availability of foods with unhealthy ingredients, and encouraging the practice of more responsible food culture. The Nourishment Concept includes 15 features.

Light Concept

The aim of the Light Concept is to minimize disruptions to the circadian rhythms of building occupants, enhance productivity, and improve physical energy and mood levels. The features in the Light Concept promote strategies for better illumination by providing criteria for window performance and design, light output and control, and appropriate visual acuity. The Light Concept includes 11 features.

Fitness Concept

The aim of the Fitness Concept is to encourage the integration of physical activity into the everyday life of building occupants. The features in the Fitness Concept promote strategies for integrating and supporting physical activity by utilizing building design, accommodating fitness regimens, and providing the space and opportunity for an active lifestyle. The Fitness Concept includes 8 features.

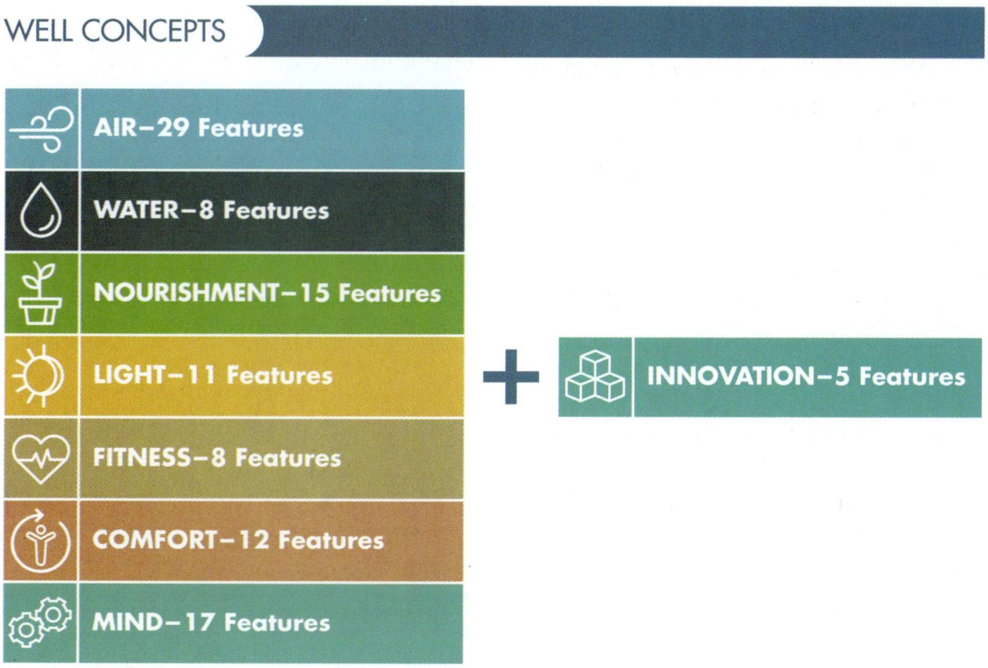

Figure 2-1. The seven WELL Concepts, as well as the Innovation features, are the factors that impact the health of building occupants.

Comfort Concept

The aim of the Comfort Concept is to design an indoor environment that is distraction-free, productive, and comfortable for building occupants. The features in the Comfort Concept promote strategies for meeting accessibility designs standards, providing comfortable furnishings and workstations, controlling acoustics and thermal conditions, and reducing known sources of discomfort. The Comfort Concept includes 12 features.

Mind Concept

The aim of the Mind Concept is to support the mental and emotional health and well-being of building occupants. The features in the Mind Concept promote strategies that provide occupants with regular feedback and knowledge about their indoor environment through design elements, relaxation spaces, and health treatments and benefits. The Mind Concept includes 17 features.

Innovation Features

The aim of the Innovation features is to promote the continuous advancement of WELL and allow project teams to achieve higher certification levels. There are five Innovation features available. Each Innovation feature counts as an optimization for any of the three project types.

WELL Features

The seven WELL Concepts are divided into features. Each feature has a specific health intent. **See Figure 2-2.** The features are further divided into parts. A *part* is a requirement of a feature that dictates the parameters or metrics to be met. Some features only contain one part, while others contain multiple parts. However, some parts are often only applicable to specific building types. Parts contain the requirements that a project must fulfill in order to achieve a feature.

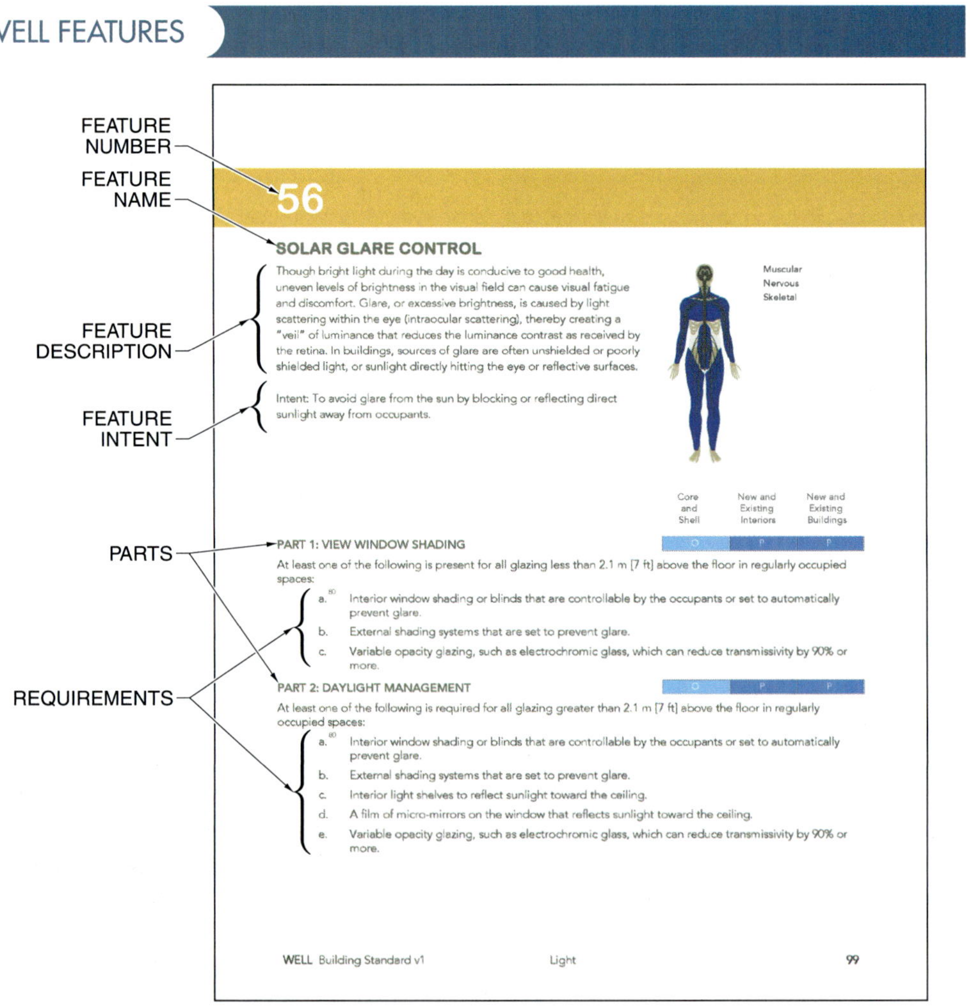

Delos Living LLC

Figure 2-2. Each of the 105 features of the WELL Building Standard contains one or more parts that are the requirements to meet that feature.

Features are classified as either preconditions or optimizations. A *precondition* is a feature that is mandatory for all levels of WELL certification. The preconditions are the core of WELL. If a project fails to meet even a single precondition, it is ineligible for even the lowest level of achievement in regards to WELL certification. Once it is ensured that a project will meet all preconditions, optimizations to achieve higher levels of certification can be pursued.

An *optimization* is an additional feature that can be used as a flexible pathway to achieve higher levels of WELL certification. Optimizations often include technologies, strategies, protocols, and designs that help a built environment, or project, achieve optimal levels of human health and well-being. It is recommended that projects be aimed to achieve as many optimizations as possible.

Synergies and Trade-offs

Synergy is the interrelationship between systems or the components of those systems that can be realized through strategic integration to achieve high levels of building performance, human performance, and environmental benefits. Many of the features in WELL were developed to have synergy with other green building and sustainability certification programs, such as the Leadership in Energy and Environmental Design™ (LEED®) green building program from the U.S. Green Building Council (USGBC) and the Living Building Challenge™ from the International Living Future Institute. **See Figure 2-3.**

Some of the features in WELL reference other green building and sustainability certification programs or provide nearly identical requirements. For example, one of the features in the Air Concept specifies that a percentage of the furniture in a project meet voluntary material reporting requirements, such as those in Imperative 10 of the Living Building Challenge 3.0. Another feature in the Air Concept has the same requirements for permanent entry walk-off systems as those found in one of the Indoor Environmental Quality (EQ) credits for the LEED v4 Building Design and Construction (BD+C) rating system. To help project teams find synergies between certification programs, the IWBI has created documents called "crosswalks," which can be found on the WELL website.

In addition, WELL Concepts also have synergy with each other. The implementation of certain strategies to achieve the requirements of one feature may also help achieve the requirements of another feature. Sometimes synergy bridges two different WELL Concepts. For example, when adjustable workstation chairs are specified to achieve ergonomic requirements for the Comfort Concept, chairs made from non-VOC-emitting materials can be specified to also increase the indoor air quality for the Air Concept.

Features within the same WELL Concept can also have synergy with each other. For example in the Light Concept, window systems can be designed to reduce energy costs while also increasing occupant wellness through adequate lighting levels and quality views of the outside.

However, there may be trade-offs between some of the WELL Concepts as well. A *trade-off* is a factoring of strategies that makes one strategy achievable while another strategy becomes too challenging or cost-prohibitive. Trade-offs between WELL Concepts and their features can affect aspects of building systems, such increased energy use or increased building costs.

SYNERGIES

Denmarsh Photography, Inc.

Figure 2-3. The Center for Sustainable Landscapes at the Phipps Conservatory and Botanical Gardens in Pittsburgh, Pennsylvania, has used synergies to become one of the greenest buildings in the world, with certifications from WELL, LEED, the Living Building Challenge, and the Sustainable SITES Initiative™.

For example, the use of reverse-osmosis (RO) filters for water purification can result in increased water use. Increases in water usage can result in higher water and sewer utility bills. Another example of a trade-off is the use of ultraviolet germicidal irradiation (UVGI) devices in a building's heating, ventilating, and cooling (HVAC) system. UVGI devices can be effective means for sanitizing air, but they also increase a building's energy load and, therefore, the energy utility bill. Project teams must carefully consider trade-offs when implementing strategies to meet the requirements of the features in the WELL Building Standard.

PROJECT TYPES

The WELL Building Standard includes three project types that are applicable to commercial and institutional buildings: New and Existing Buildings, New and Existing Interiors, and Core and Shell. A project type encompasses the entire scope of a project. The features in the WELL Concepts that are applicable to a project depend on the project's type. All project types can also achieve up to five additional optimizations through the Innovation features.

New and Existing Buildings

For maximum effects on wellness and human health, WELL is best applied to entire buildings. The New and Existing Buildings project type addresses the full scope of building design and construction, as well as aspects of building operations. The New and Existing Buildings project type can only be applied to buildings in which a minimum of 90% of the total floor area is occupied by the building owner and is operated by the same management. There are 41 preconditions and 59 optimizations (100 total features) available to New and Existing Buildings in the WELL Building Standard.

New and Existing Interiors

WELL can also be applied to the interior of a building for the wellness and human health of building occupants. The New and Existing Interiors project type addresses portions or the entirety of a building's interior space, unless the building is undergoing major renovations. There are 36 preconditions and 62 optimizations (98 total features) available to New and Existing Interiors in the WELL Building Standard.

Core and Shell

For the wellness and health of future building occupants, WELL can also be applied to the exterior shell and core mechanical and plumbing systems of a building. The Core and Shell project type addresses the basic structure and shape of a building as well as its heating, ventilating, and air conditioning (HVAC) systems and its water quality. Many of the wellness goals for Core and Shell projects can help streamline certification for New and Existing Interiors if it is pursued later.

The Core and Shell project type applies to projects where up to 25% of the project area is fully controlled by the building owner. At least 75% of the project space must be occupied by one or more tenants. However, all of the building space must adhere to the Core and Shell requirements. There are 26 preconditions and 28 optimizations (54 total features) available to Core and Shell projects in the WELL Building Standard.

Pilot Standard Programs

Although the features in the WELL Building Standard can be applied to many types of commercial and institutional buildings, there are other building types and spaces that can also benefit from WELL. These different types of buildings and spaces may have unique needs and challenges that require specific solutions. For this reason, a pilot standard program has been created by the IWBI.

Currently, there are six WELL pilot programs available to projects that meet specific space types. These pilot programs allow IWBI to gather data on new innovations for future versions of WELL. The pilot programs attempt to address the specific needs of different types of buildings as well as test the applicability of WELL for different real-estate sectors. This allows the WELL Building Standard to continually evolve and new project types to be added. The following pilot programs are available to projects that meet the space type:

- **Multifamily residential.** The Multifamily Residential pilot program only applies to single-building projects that have at least five dwelling units, such as apartments, condominiums, and townhouses.
- **Educational facilities.** The Educational Facilities pilot program applies to places of learning where dedicated staff are employed for the purpose of teaching any age or grade level, including elementary schools, middle schools, high schools, and higher-education facilities.
- **Retail.** The Retail pilot program applies to commercial retail locations, including boutique shops, stand-alone big box stores, and shopping malls or the individual retail spaces within them, where consumers are considered transient occupants but the staff are regular building occupants.
- **Restaurant.** The Restaurant pilot program only applies to staffed or self-serve eating spaces where a consumer purchases and dines on-site, either indoors or outdoors (neither kitchens nor take-out-only locations or locations where the primary source of revenue comes from the sale of alcoholic beverages are covered by this pilot program).
- **Commercial kitchens.** The Commercial Kitchens pilot program only applies to spaces where cooks prepare food for other building occupants or visitors and must be paired with another pilot program or one of the three main project types in the WELL Building Standard.
- **Communities.** The Communities pilot program applies to community projects that meet minimum occupancy, building size, and use requirements. These requirements include the use of integrated, mixed-use developments. Any Communities pilot project must contain at least one building that achieves certification under a qualifying health and wellness standard.

Space Types. IWBI uses space types to define a building or a part of a building by its specific use or function for the purposes of certification through the pilot programs. The two space types used to define a building or a part of a building are primary and secondary spaces. A primary space that applies to an entire building can be certified under the pilot program that matches that primary space. A secondary space must be paired with a primary space for certification. For example, project team seeking Commercial Kitchen certification for a school cafeteria must also seek certification for the entire school building through the Educational Facilities pilot program.

HUMAN HEALTH AND THE BUILT ENVIRONMENT

Each feature in the WELL Building Standard affects one or more of a human's body systems. **See Figure 2-4.** Much of human health depends on the health of these various body systems. The creators of the WELL Building Standard drew on scientific studies and medical research to integrate building design and policy strategies that protect or increase the health of body systems. The features in the WELL Concepts can be used to protect as well as lessen the negative effects of the built environment on body systems such as the cardiovascular, digestive, endocrine, immune, integumentary, muscular, nervous, reproductive, respiratory, skeletal, and urinary systems.

Cardiovascular System

The cardiovascular system, which consists of the heart, blood vessels, and blood, primarily supplies nutrients, carries oxygen and carbon dioxide, and removes waste from body tissues. WELL aims to maintain or increase cardiovascular health in the built environment by mitigating stress, encouraging healthy nutrition, providing opportunities for physical activity, and eliminating environmental health hazards in the air and water.

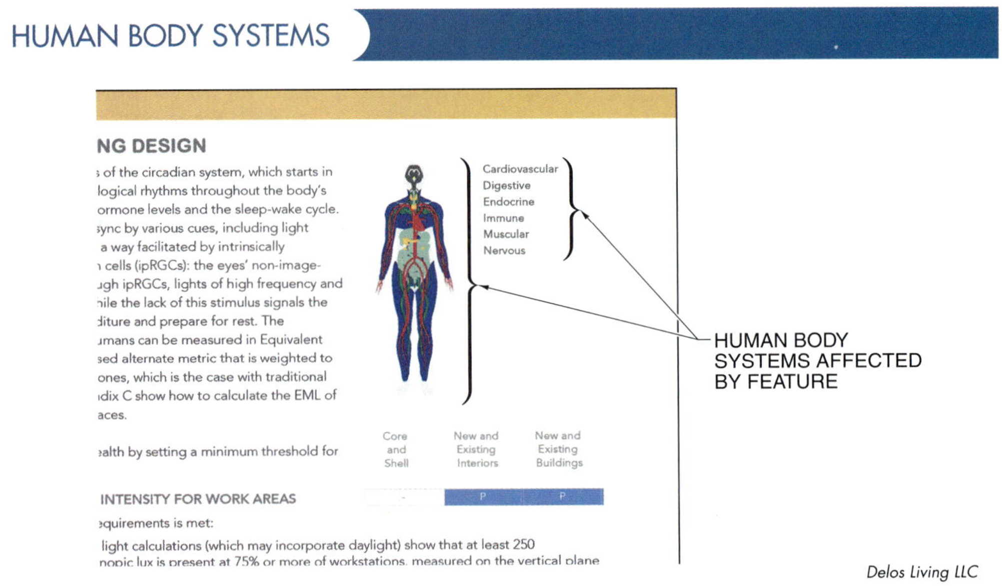

Figure 2-4. The body systems affected by the requirements for each feature are listed in the WELL Building Standard.

Digestive System

The digestive system, which consists of the mouth, esophagus, stomach, small and large intestines, liver, and pancreas, produces digestive hormones and enzymes and breaks down food for the absorption and use of nutrients. WELL aims to support proper digestive health by mitigating stress and encouraging healthy nutrition.

Endocrine System

The endocrine system, which consists of hormone-secreting glands, produces chemical compounds to regulate many bodily processes, such as growth, immunity, metabolism, reproduction, mood, and digestion. WELL aims to maintain healthy endocrine functions by mitigating stress, encouraging healthy nutrition, and eliminating environmental health hazards in the air and water.

Immune System

The immune system, which consists of highly specialized cells, proteins, tissues, and organs, acts as the body's defense against internal and foreign disease-causing organisms and substances. WELL aims to promote and enhance the health of the immune system by eliminating environmental health hazards in the air and water, mitigating stress, encouraging healthy nutrition, and providing opportunities for physical activity.

Integumentary System

The integumentary system, which consists of the skin, hair, and nails, protects internal organs, prevents water loss, regulates body temperature, and protects the body against foreign pathogens and harmful toxins. WELL aims to maintain the integrity of the integumentary system by encouraging healthy nutrition and eliminating environmental health hazards.

Muscular System

The muscular system, which consists of skeletal, smooth (involuntary), and cardiac muscles, supports posture, physical movement, blood circulation, and digestion. WELL aims to promote building design and furnishing strategies that reduce injury to the muscular system and strengthen muscles at the same time by providing opportunities for physical activity and supporting proper ergonomics.

Nervous System

The nervous system, which consists of the brain, spinal cord, and nerves, directly and indirectly controls bodily processes, including movement, cognition, and vital organ functions. WELL aims to support cognitive functions and protect the health of the nervous system by mitigating stress, encouraging healthy nutrition, providing opportunities for physical activity, eliminating environmental health hazards in the air and water, and promoting quality sleep.

Reproductive System

The reproductive system, which consists of hormone-secreting glands and the reproductive organs, supports the biological process of reproduction. WELL aims to protect reproductive health by encouraging healthy nutrition, providing opportunities for physical activity, and eliminating environmental health hazards in the air and water.

Respiratory System

The respiratory system, which consists of the mouth, nose, diaphragm, trachea, bronchi and bronchioles, and the lungs, works with the cardiovascular system to provide oxygen to the body and remove carbon dioxide. WELL aims to optimize the function and health of the respiratory system by eliminating airborne environmental health hazards and providing opportunities for physical activity to strengthen the lungs.

Skeletal System

The skeletal system, which consists of bones, marrow, cartilage, tendons, and ligaments, protects internal organs, stores minerals, produces blood cells, and aids in hormone regulation. WELL aims to support the health and function of the skeletal system by encouraging healthy nutrition, providing opportunities for physical activity, and supporting proper ergonomics.

Urinary System

The urinary system, which consists of the kidneys, ureters, bladder, and urethra, filters toxins, balances pH levels in the blood, helps maintain blood pressure, and eliminates waste from the body through urination. WELL aims to protect urinary health by mitigating stress and eliminating waterborne environmental health hazards.

KEY TERMS AND DEFINITIONS

feature: One of the 105 sections of the WELL Building Standard with a specific health intent.

International WELL Building Institute (IWBI): A public-benefit organization that administers the WELL Building Standard and aims to improve human health and well-being through the built environment.

optimization: An additional feature that can be used as a flexible pathway to achieve higher levels of WELL certification.

part: A requirement of a feature that dictates the parameters or metrics to be met.

precondition: A feature that is mandatory for all levels of WELL certification.

KEY TERMS AND DEFINITIONS *(continued)*

synergy: The interrelationship between systems or the components of those systems that can be realized through strategic integration to achieve high levels of building performance, human performance, and environmental benefits.

trade-off: A factoring of strategies that makes one strategy achievable while another strategy becomes too challenging or cost-prohibitive.

WELL Building Standard: A performance-based system for measuring, certifying, and monitoring features of the built environment that impact human health and well-being through air, water, nourishment, light, fitness, comfort, and mind. Simply referred to as WELL.

CHAPTER REVIEW

Completion

______ **1.** The WELL Building Standard is organized into ___ major categories of wellness called concepts.

______ **2.** A feature is one of the ___ sections of the WELL Building Standard with a specific health intent.

______ **3.** A(n) ___ is a requirement of a feature that dictates certain parameters or metrics to be met.

______ **4.** The aim of the ___ is to promote the continuous advancement of WELL and allow project teams to achieve higher certification levels.

______ **5.** A(n) ___ is a feature that is mandatory for all levels of WELL certification.

______ **6.** A(n) ___ is an additional feature that can be used as a flexible pathway to achieve higher levels or WELL certification.

______ **7.** Many of the features in WELL were developed to have ___ with other green building and sustainability programs, such as the Leadership in Energy and Environmental Design™ (LEED®) green building program from the U.S. Green Building Council (USGBC) and the Living Building Challenge™ from the International Living Future Institute.

______ **8.** ___ between WELL Concepts and their features can affect aspects of building systems, such as increased energy use or increased building costs.

______ **9.** The New and Existing Buildings project type can only be applied to buildings where a minimum of ___% of the total floor area is occupied by the building owner and is operated by the same management.

______ **10.** The ___ project type addresses portions or the entirety of a building's interior space, unless the building is undergoing major renovations.

CHAPTER REVIEW *(continued)*

____________________ **11.** The Core and Shell project type can only be applied to buildings where up to ___% of the space is fully controlled by the building owner.

____________________ **12.** Currently, there are ___ WELL pilot programs available to projects that meet specific space types.

Short Answer

1. List the number of features within each of the seven WELL Concepts, as well as the number of Innovation features.

2. What are two examples of synergy between the WELL Building Standard and other green building and sustainability certification programs?

3. What is an example of synergy between the Comfort Concept and the Air Concept?

4. List the three project types of the WELL Building Standard and the number of preconditions and optimizations that apply to each.

5. List the six current pilot standard programs for the WELL Building Standard.

6. What are the human body systems that the features of the WELL Concepts seek to protect by lessening the negative effects of the built environment?

WELL AP EXAM PRACTICE QUESTIONS

1. How many features are included in the Water Concept?
   A. 8
   B. 12
   C. 15
   D. 29

2. A project team wishes to provide building occupants with the space and opportunity for an active lifestyle as well as encourage healthy eating habits that will enable that lifestyle. Which two WELL Concepts may have synergies between their features that the project team can use to implement this plan?
   A. Nourishment and Mind
   B. Fitness and Comfort
   C. Nourishment and Fitness
   D. Fitness and Mind

3. The WELL project team for a soon-to-be constructed office building is looking at registering the project for the New and Existing Building project type. What does this project type encompass?
   A. The design of the building, but not the construction
   B. Major renovations of existing spaces
   C. The design and installation of the mechanical systems and window locations and glazing
   D. The full scope of project design and construction as well as aspects of building operations

4. What is the minimum percentage of project space that must be occupied by one or more tenants in order for a project to qualify for the Core and Shell project type?
   A. 10%
   B. 25%
   C. 50%
   D. 75%

5. Which WELL Concept aims to minimize disruptions to the circadian rhythms of building occupants, enhance productivity, and improve physical energy and mood levels?
   A. Air
   B. Light
   C. Fitness
   D. Mind

CHAPTER THREE

THE WELL CERTIFICATION PROCESS

Because of a WELL AP's extensive knowledge of the WELL Building Standard and certification process, the WELL AP working on a specific project will lead the coordinated effort to achieve certification. The WELL AP will ensure that the WELL certification process runs smoothly by ensuring that every team member understands the steps in the process and their role in achieving those milestones.

OBJECTIVES

- Explain how WELL certification supports the triple bottom line and increases project value.
- Identify the three levels of achievement for WELL certification.
- Describe the timeline for WELL certification.
- Identify the members of a project team and their responsibilities.
- Explain the role of a WELL assessor in the certification process.
- Explain the types of documentation that must be submitted and the methods of performance verification that are necessary for WELL certification.
- Describe the WELL report that is received for a project.
- Explain the purpose of alternative adherence paths (AAPs).
- Describe how a curative action plan can be submitted and an appeal can be filed.
- Explain the process for recertification.
- Describe the key differences in project certification for the WELL Community Standard compared to certification under the WELL Building Standard.

KEY TERMS

- alternate adherence path (AAP)
- appeal
- charrette
- curative action plan
- owner
- performance verification
- project administrator
- stakeholder
- triple bottom line
- WELL Accredited Professional (WELL AP)
- WELL assessor
- WELL report

Learner Resources
ATPeResources.com/QuickLinks
Access Code: 935534

WELL CERTIFICATION

A project that has received WELL certification helps create a healthy built environment for both its occupants and visitors. The WELL certification process takes a human-centered approach to building design and construction; however, achieving certification can result in benefits other than just a healthy building. Projects that have successfully achieved certification through the WELL Building Standard provide value in many different aspects, including the triple bottom line and the business aspect.

WELL and the Triple Bottom Line

The choice to pursue WELL certification requires the evaluation of the project's triple bottom line. The *triple bottom line* is the concept of sustainability that includes the financial, environmental, and social bottom lines of a project. **See Figure 3-1.**

TRIPLE BOTTOM LINE

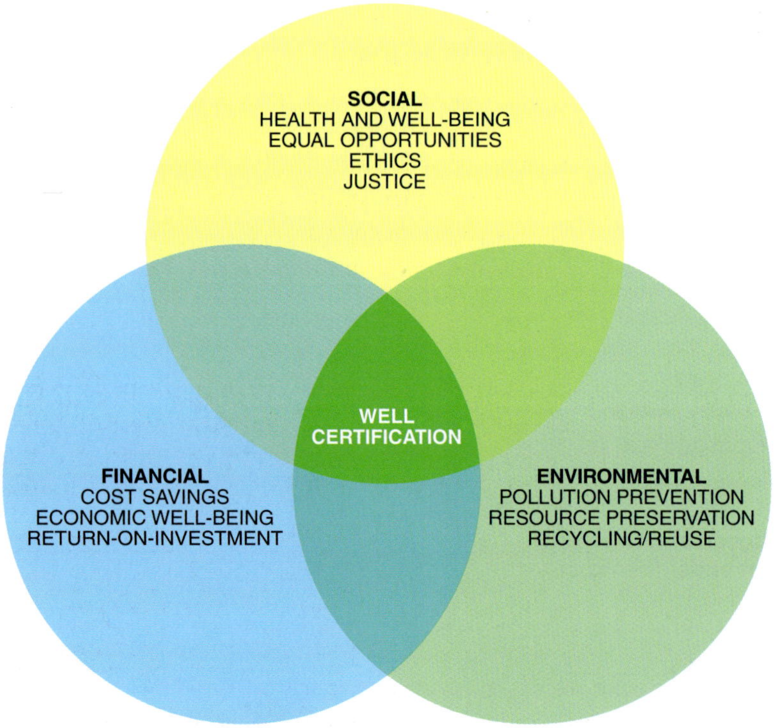

Figure 3-1. The triple bottom line is the concept of sustainability that involves the financial, social, and environmental bottom lines of a project.

The financial bottom line focuses on the economic well-being of a company, how the project is financed, and the return-on-investment (ROI) that the project provides. The environmental bottom line addresses the project's impact on the land, air, water, plants, and animals located around the project site as well as the regional and global environmental concerns of the project. The social bottom line focuses on the impacts of the project during construction and into its occupancy on the happiness, health, and productivity of the project stakeholders.

A *stakeholder* is anyone who is affected, or will be affected, by the construction and operation of a building, such as the project owners, company shareholders, building occupants, visitors, construction and maintenance personnel, and even the surrounding community. Collaboration among stakeholders is an integral aspect of the construction and design of healthy built environments.

The WELL Building Standard is primarily focused on the social bottom line to address the health and wellness of project stakeholders. However, the financial bottom line may be positively or negatively affected as WELL requirements are met. Although WELL does not focus on the environmental bottom line, benefits may be found when a project's environmental impact is considered during project development, especially if other green building certifications are sought.

Business Case for WELL

WELL certification also provides value in a business sense. A project that has successfully gone through the WELL certification process may be perceived as having added value. Buildings that have met rigorous green building standards, such as the WELL Building Standard, can be used for marketing and branding opportunities. Since WELL focuses on the health and wellness of the people working in and around a project, there may also be increases in productivity and employee retention. These increases can lead to higher profits. In addition, healthier employees can reduce the health insurance costs for the company and the employees.

CERTIFICATION PROCESS

Every project that aims to achieve certification in the WELL Building Standard must go through the WELL certification process. Through the WELL certification process, it can be determined whether a project can be WELL Certified™ by assessing and verifying the project's adherence to every applicable requirement of each pursued feature. Verification is necessary for the intents of the WELL Building Standard because some features allow flexibility in achieving their performance-based requirements. Other features call for prescriptive solutions, such as specific technologies, design strategies, or policies, which must be documented as being implemented or achieved.

The WELL Building Standard has three levels of achievement for certification. Projects must achieve all applicable preconditions for each project type as well as certain percentages of optimizations to become WELL Certified. **See Figure 3-2.** Projects seeking certification for all project types or the pilot standards can be certified to the Silver, Gold, or Platinum levels, which require increasing percentages of achieved optimization.

A project is WELL Certified™ Silver when it meets all of the applicable preconditions but no optimizations. A project is WELL Certified™ Gold when it meets all of the applicable preconditions and 40% or more of the applicable optimizations. A project is WELL Certified™ Platinum when it achieves all of the applicable preconditions and 80% or more of the applicable optimizations.

WELL Certification Steps

The length of time for WELL certification varies depending on certain variables like the scope of the project, the WELL certification level targeted, and the thoroughness of the WELL project team. However, the basic WELL certification process can be broken down into five general steps.

1. Registration. Projects aiming for one of the certification levels of the WELL Building Standard must first be registered through WELL Online. WELL Online is the web-based portal used to manage and complete the WELL certification process as well as register for the WELL AP exam. Once a project is registered, a WELL assessor will be assigned to it. The project team can also register and view an orientation webinar through WELL Online.

2. Documentation review. The second step of the WELL certification process is to submit the necessary documentation, such as annotated project documents, drawings, and letters of assurance, through WELL Online. The documentation will undergo a technical review performed by the WELL assessor. All documentation must be approved before the project may move on to the next step of performance verification.

3. Performance verification. Performance verification is an integral part of the WELL Building Standard. This step of the certification process includes a site visit from the WELL assessor to ensure that the project performs as intended. After this step is complete, a project is issued a WELL report.

4. Certification. Once a WELL report has been accepted as final and all applicable requirements are met, the project is recognized as WELL compliant or WELL Certified. The project will receive the official award, a plaque showing its level of certification achievement, and other promotional and informational documents.

5. Recertification and documentation submission. WELL recertification is required every three years. Recertification ensures that a project continues to perform as intended by requiring another round of documentation submission and performance verification.

WELL Building Standard Levels of Achievement

Version of Standard	Certification Level	Preconditions That Must Be Achieved	Optimizations That Must Be Achieved
WELL Building Standard	Silver	All applicable	None
	Gold	All applicable	40% of applicable
	Platinum	All applicable	80% of applicable
WELL Pilot Standard Programs	Silver	All applicable	20% of applicable
	Gold	All applicable	40% of applicable
	Platinum	All applicable	80% of applicable

Figure 3-2. Projects must achieve all applicable preconditions for each project type as well as certain percentages of optimizations in each WELL Concept to become WELL Certified™.

Project Size

During registration, the project size (gross area) in square feet or square meters must be provided. The gross square footage (square meters) for a project includes the sum of the floor areas of the spaces within the building. Measurements must be taken from the exterior faces of the exterior walls or from the centerline of walls separating buildings. The calculations must include basements, mezzanines and intermediate-floored tiers, and penthouses, if they have a headroom height of 7.5′ or greater.

However, the calculated gross square footage should not include nonenclosed (or nonenclosable) roofed-over areas, such as exterior covered walkways, porches, terraces or steps, roof overhangs, and similar features. Nor should the calculations include air shafts, pipe trenches, chimneys, or floor areas dedicated to parking and motor vehicle circulation.

Project Teams

The WELL project team is a group of people from various entities who are tasked to perform the actions necessary for WELL certification. Each member of the project team has particular responsibilities, which include submitting all required documents.

Project team members should meet for charrettes during all phases of the construction and design of a building seeking WELL certification. A *charrette* is an intensive, multiparty workshop that brings people together to explore, generate, and collaboratively produce building design options. Team members must also be available during performance verification. Project teams consist of a project administrator, the project owner, the WELL Accredited Professional (WELL AP), and other licensed professionals.

Project Administrator. A *project administrator* is the individual who acts as project manager and oversees the WELL certification process. The project administrator is responsible for checking and submitting complete and accurate documentation for review through WELL Online. The role of project administrator can be assigned to a WELL AP, the building owner, or another member of the project team.

Owner. The *owner* is an individual property owner, or a representative of the entity that owns the property, who is responsible for authorizing the registration of a project on WELL Online. Owners have the authority to make decisions concerning the property on which a project is being built. If multiple individuals have ownership rights, a single person must be selected as the sole authorized decision-maker for the purposes of WELL certification.

WELL Accredited Professional. A *WELL Accredited Professional (WELL AP)* is an individual who possesses the knowledge and skill necessary to support the WELL certification process. A WELL AP is included on a project team as a consultant who can guide successful WELL certification. Although it is not required to have a WELL AP on the project team, the presence of a credentialed professional will help ensure the successful implementation of WELL features and WELL certification.

Licensed Professionals. For WELL certification, letters of assurance are often required from licensed professionals as documentation that requirements have been met. These individuals are often included on the project team. Licensed professionals typically found on project teams include architects, contractors, interior designers, safety/environmental compliance officers, wellness coordinators, acoustical consultants, and mechanical, electrical, and plumbing (MEP) engineers.

WELL Assessors

Green Business Certification Inc. (GBCI) assigns a WELL assessor to every project after document submission. A *WELL assessor* is an independent third party who conducts on-site performance tests, inspections, and spot checks, as well as reviews documentation in order to evaluate a project's eligibility for WELL certification. A WELL assessor works closely with the project team. **See Figure 3-3.** WELL assessors are not the same individual as the WELL AP on the project team, though the WELL assessor may hold a WELL AP credential.

WELL ASSESSOR CERTIFICATION ROLES

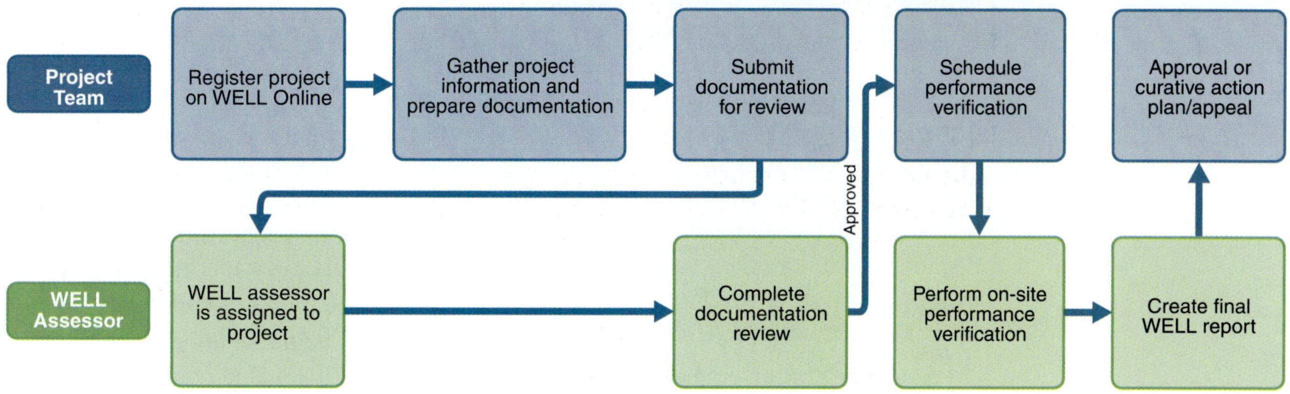

Figure 3-3. A WELL assessor conducts on-site performance tests, inspections, and spot checks, as well as reviews documentation in order to evaluate a project's eligibility for WELL certification.

One of the roles of a WELL assessor is to provide technical support to a project team seeking WELL certification. A WELL assessor can clarify WELL requirements for a project team that cannot find the answers it needs in published documents. In addition, if a project team encounters a unique or difficult scenario during any of the project phases, it can seek the help of the WELL assessor.

The primary role of a WELL assessor is to evaluate whether a project pursuing certification meets the requirements of the WELL Building Standard. During this essential task, the WELL assessor reviews the project's documentation and executes the on-site performance verification to verify that the project performs as intended. If adherence to the WELL Building Standard is verified, the WELL assessor creates the WELL report and determines the level of certification achieved.

DOCUMENTATION AND PERFORMANCE VERIFICATION

Like WELL APs, WELL assessors must be knowledgeable in the WELL Building Standard and have experience in the building industry due to the wide range of documentation and performance verification tasks that they must conduct. Each WELL feature is achieved through documentation, performance verification, or a combination of both. WELL is a performance-based standard, and it is the job of a WELL assessor to verify that the features implemented in a project are documented and perform as intended.

Types of Documentation and Performance Verification

When a project seeks WELL certification, its adherence to each part of the features it intends to earn may require different types of documentation or performance verification. All documentation must be submitted before performance verification can begin. The types of documentation that are used to verify whether a project meets specific requirements of the WELL Building Standard include letters of assurance, annotated documents, and general documents. The methods of performance verification include different types of on-site checks.

Letters of Assurance. Some features may require a separate letter of assurance from the licensed professional who is overseeing the implementation of a specific feature. Letters of assurance may be required from architects, contractors, or MEP engineers. Letters of assurance are submitted via WELL Online by the project administrator or other project team member and then reviewed by the WELL assessor.

Annotated Documents. Some features may require that existing documents, such as those associated with the construction of a project, be annotated to prove how the requirements of a specific feature are met. Annotated documents include design drawings with highlighted information, operations schedules, policy documents and employee handbooks, and project narratives and balancing reports. Annotated documents are submitted via WELL Online by the project administrator or other project team member and then reviewed by the WELL assessor.

General Documents. Some documents are required for the certification process that are not tied to a specific part or feature. These documents are used to inform IWBI and the WELL assessor of the general project details. General documents do not require annotation and are used for the certification process at large. Some required general documents are representative floor plans or project maps, mechanical drawings, and narratives describing the project in general and listing the attempted features.

Performance Verification. *Performance verification* is a site visit in which a WELL assessor conducts performance tests, visual inspections, and spot checks, and that also includes follow-up analyses of collected data and samples from the site. Performance verification is required for many features in the WELL Concepts. However, the WELL assessor can also choose to verify the performance of any feature pursued by the project, including those for which required documentation has already been submitted. The reason for a WELL assessor to conduct on-site assessments on already documented features is to confirm that the requirements of those features are satisfied.

The WELL assessor must visit the project site in order for the performance verification requirements to be fulfilled. Depending on the project type or pilot program, certain conditions must be met before performance verification can be scheduled. **See Figure 3-4.** For example, all the necessary documentation must already be submitted, reviewed, and approved through WELL Online. Most projects must also have been issued a certificate of occupancy at least one month before the performance verification. Some projects must have a minimum of 50% of the expected occupancy.

During performance verification, the WELL assessor will conduct performance tests, visual inspections, spot checks, and spot measurements. This process may take several days. Additional time may be required for third-party laboratories to analyze samples that were taken by the WELL assessor per the IWBI's sampling protocols. The environmental properties that a WELL assessor may test on-site include air quality, water quality, lighting quality, thermal conditions, and acoustics.

WELL Reports

After the WELL assessor has visited the project site and completed performance verification, a project's WELL report will be available on WELL Online in about 40 to 45 business days. A *WELL report* is a comprehensive report of a project that includes a feature-by-feature summary indicating whether the project team successfully provided documentation to verify the achievement of each feature or whether the project has successfully met the measurable criteria for specific features.

Alternative Adherence Paths

Besides the Innovation features in the WELL Building Standard, Features 101 through 105, another way for a project to pursue creative solutions is through alternative adherence paths. An *alternative adherence path (AAP)* is an alternative solution for meeting the intent of a WELL feature requirement. A project team may submit as many AAPs as it wishes, but each request can only be applied to a single feature. The only other requirement for an AAP is that the request must be supported by sound medical, scientific, or industry research.

Project teams may submit AAPs for preapproval before or after registration. Submission forms are available on the WELL website. Fees do not apply for the first three AAPs, but a fee is required for each additional AAP. Pilot projects receive 10 free AAP submissions. AAPs that are approved and have broad applicability may be incorporated into future versions of the WELL Building Standard as official pathways.

A single AAP may be submitted and applied to multiple projects from the same entity. All of the projects must be located at the same development site, or the submitted AAP must be for a feature categorized as a WELL organizational protocol. The policy or schedule requirements in a WELL organizational protocol feature (Features 65, 66, 90, 91, 92, 93, 94, 95, 96, and 98) must be adopted by the entire entity seeking certification in the project's building and must be available independent of an occupant's location within the building.

Also, the proposed path must be identical across all projects using the AAP. The project team must specify the names and IDs of all projects using the AAP in the AAP submission. Only projects registered for WELL certification may be specified.

Curative Actions and Appeals

Once a WELL assessor issues the WELL report, the project owner has 180 calendar days to either accept the report as is, initiate a curative action plan to address unmet criteria, or appeal the findings in the report. If the report is accepted as is, then the project will be awarded the level of certification or compliance dictated on the WELL report. Fees apply for both the curative action plan and appeal as well as any additional testing or retesting that may be involved.

Curative Action Plans. A *curative action plan* is a document that outlines strategies that a project team will employ to address unmet criteria as identified in the WELL report. The curative action plan must contain a clear, specific, and feasible plan to address the unmet criteria. The plan is reviewed by the project's WELL assessor, and results are provided within 20 to 25 business days. Once approved, the plan can be implemented. Once the curative action plan is completed, follow-up performance verification can take place. This results in the issuance of a new WELL report that supersedes any previous reports.

Appeals. Project teams have the right to appeal the findings of a WELL assessor or IWBI. An appeal is a document that outlines a project team's disagreement with any finding of the WELL report or any decision regarding proposals for AAPs, curative actions, or the Innovation features. Each feature in dispute must have its own letter of appeal. An appeal must contain an explanation of the basis for the appeal and identify suspected errors. Appeals for the WELL report must be submitted within 180 days of the report being issued, while appeals for other decisions must be submitted within 90 days. The GBCI will respond within 20 to 25 business days with an appeal review report. The project team may either accept or reject the report, but there can only be one subsequent appeal for each feature. The subsequent appeals must be submitted within 90 days of the issuance of the appeal review report being challenged.

Performance Verification Requirements

Type/Program	Documentation Approved	One Month from Certificate of Occupancy	50% Occupancy
New and Existing Buildings New and Existing Interiors	Yes	Yes	Yes
Core and Shell	Yes	No	No
Retail Pilot Program Educational Facilities Pilot Program Commercial Kitchens Pilot Program	Yes	Yes	No
Multifamily Residential Pilot Program	Yes	No	No

Figure 3-4. Depending on the project type or pilot program, certain conditions must be met before performance verification can be scheduled.

PRECERTIFICATION

Precertification, an optional review pathway available to all project types, may be used by project teams to determine which features the project is likely to achieve during the full WELL certification review process. Precertification focuses on the intended design, construction, and operational strategies for the project. The process for precertification involves the completion of the WELL Precertification Worksheet, which is then submitted through WELL Online. The project's WELL assessor reviews the worksheet and provides a report. If all the requirements are met, the project receives the WELL Precertification level and certificate award. An additional fee applies to the precertification process.

CONTINUED ENGAGEMENT

During the time after WELL certification and before recertification, the project must address specific ongoing requirements depending on the feature requirements achieved. Some of these may include post-occupancy survey results, proof of maintenance, and ongoing environmental parameter measurements. These features generally require documentation to be submitted to the GBCI for review on an annual basis. To remain in good standing, projects must submit the documents within 15 months of certification and then every 12 months thereafter. Failure to provide these documents within this time frame may result in an additional fee assessed at recertification. If the ongoing records are not submitted at all by recertification, then the project will not only be assessed the additional fee but will also be ineligible to pursue these features for the following certification period.

RECERTIFICATION

Registration for recertification for WELL projects is mandatory within three years (36 months) of the initial certification to ensure that the projects continue to perform according to the requirements of the WELL Building Standard. If a project team fails to apply for recertification for a project through WELL Online before three years has passed, the project's certification will expire. If data for achieved features is required to be submitted annually to IWBI, the project team must still submit or report the data during these three years.

For recertification, buildings must again undergo performance verification according to the version of the WELL Building Standard originally used for certification or an updated version of the project team's choosing. Recertification can positively or negatively affect a project's level of certification. If a project has failed to upkeep its WELL features, it may even lose its certification. However, a project can increase its level of achievement and receive a higher level of certification. The recertification process must be completed within six months (42 months after initial certification) after the project team has submitted the recertification application.

WELL COMMUNITY STANDARD CERTIFICATION

Projects seeking certification for the Communities pilot program use the WELL Community Standard™. The WELL Community Standard was developed in a manner similar to the WELL Building Standard, through research and development by leading professionals in medicine, science, and the green building and construction industry. However, because the WELL Community Standard expands the scope of the standard from a single building to an entire community or district, some aspects of its certification process differ from certification under the WELL Building Standard. Some of the key differences include the following:

- The project size for projects seeking certification under the WELL Community Standard should include the gross area of all land and water areas, both owned and not owned by the project owner, within the project boundary.

- For the WELL Community Standard, the owner is the individual or organization that registers the project, pays the applicable fees, and is authorized to fulfill the obligations of that pilot standard.
- WELL Community Standard projects must complete documentation submission within 5 years of registration and undertake and submit measurements for performance verification within 10 years of registration.
- Precertification is not available for projects pursuing the WELL Community Standard. WELL Community Standard projects should use the Design Verified designation prior to completion of construction. The IWBI awards WELL Community Standard projects with this designation when all documents have been reviewed and approved. The Design Verified designation allows projects to market the proposed wellness features and communicate their progress toward WELL certification.
- WELL certification for the WELL Community Standard is valid for 5 years. If a WELL Community Standard project does not file an application for recertification before the expiration of the original certification period or fails to successfully obtain recertification within 66 months of receiving the initial certification, the project's WELL certification will expire.
- WELL Community Standard projects may change their project boundary at the time of recertification. The new project will then be evaluated as a whole under the new boundary.

KEY TERMS AND DEFINITIONS

alternative adherence path (AAP): An alternative solution for meeting the intent of a WELL feature requirement.

appeal: A document that outlines a project team's disagreement with any finding of the WELL report or any decision regarding proposals for AAPs, curative actions, or the Innovation features.

charrette: An intensive, multiparty workshop that brings people together to explore, generate, and collaboratively produce building design options.

curative action plan: A document that outlines strategies that a project team will employ to address unmet criteria as identified in the WELL report.

owner: An individual property owner, or a representative of the entity that owns the property, who is responsible for authorizing the registration of a project on WELL Online.

performance verification: A site visit in which a WELL assessor conducts performance tests, visual inspections, and spot-checks, and that also includes follow-up analyses of collected data and samples from the site.

project administrator: The individual who acts as project manager and oversees the WELL certification process.

stakeholder: Anyone who is affected, or will be affected, by the construction and operation of a building, such as the project owners, company shareholders, building occupants, visitors, construction and maintenance personnel, and even the surrounding community.

triple bottom line: The concept of sustainability that includes the financial, environmental, and social bottom lines of a project.

WELL Accredited Professional (WELL AP): An individual who possesses the knowledge and skill necessary to support the WELL certification process.

WELL assessor: An independent third party who conducts on-site performance tests, inspections, and spot checks, as well as reviews documentation in order to evaluate a project's eligibility for WELL certification.

WELL report: A comprehensive report of a project that includes a feature-by-feature summary indicating whether the project team successfully provided documentation to verify the achievement of each feature or whether the project has successfully met the measurable criteria for specific features.

CHAPTER REVIEW

Completion

______ 1. A(n) ___ is anyone who is affected, or will be affected, by the construction and operation of a building, such as the project owners, company shareholders, building occupants, visitors, construction and maintenance personnel, and even the surrounding community.

______ 2. A project is WELL Certified™ ___ if it meets all applicable preconditions and less than 40% of the optimizations.

______ 3. A(n) ___ is an intensive, multiparty workshop that brings people together to explore, generate, and collaboratively produce building design options.

______ 4. A(n) ___ is the individual who acts as project manager and oversees the WELL certification process.

______ 5. The ___ is an individual property owner, or a representative of the entity that owns the property, who is responsible for authorizing the registration of the project on WELL Online.

______ 6. A(n) ___ is an independent professional who conducts on-site performance tests, inspections, and spot checks, as well as reviews documentation in order to evaluate a project's eligibility for WELL certification.

______ 7. WELL assessors are not the same individual as the ___ on the project team.

______ 8. Some features may require a separate ___ from the licensed professional who is overseeing the implementation of a specific feature.

______ 9. ___ include design drawings with highlighted information, operations schedules with time logs, policy documents and employee handbooks, and project narratives and balancing reports.

______ 10. During ___, the WELL assessor will conduct performance tests, visual inspections, and spot checks.

______ 11. A(n) ___ is an alternative solution for meeting the intent of a WELL feature.

______ 12. Once a WELL assessor issues the WELL report, the project owner has ___ calendar days to either accept the report as is, initiate a curative action plan to address unmet criteria, or appeal the findings in the report.

______ 13. A(n) ___ is a document that outlines strategies that the project team will employ to address unmet criteria as identified in the WELL report.

______ 14. A(n) ___ is a document that outlines a project team's disagreement with any finding of the WELL report or any decision regarding proposals for alternative adherence paths, curative actions, or the Innovation features.

______ 15. Recertification for the New and Existing Interiors, New and Existing Buildings, and Core and Shell project types is mandatory within ___ years to ensure that these projects continue to perform according to the requirements of the WELL Building Standard.

______ 16. Projects seeking certification for the Communities pilot program use the ___.

______ 17. Because precertification is not available to WELL Community Standard projects, these projects should use the ___ designation that is awarded by the IWBI when all documents have been reviewed and approved.

CHAPTER REVIEW *(continued)*

Short Answer

1. List the three components that compose the triple bottom line.

2. How does WELL certification provide value?

3. What are the levels of certification and the requirements in terms of preconditions and optimizations that can be achieved for the New and Existing Interiors, New and Existing Buildings, and Core and Shell project types?

4. List the five basic steps of the WELL certification process.

5. List the types of licensed professionals typically found on the project team for WELL certification.

6. What are the primary duties of a WELL assessor in the WELL certification process?

7. What is included in the WELL report that a project team receives after the WELL assessor has visited the project site and completed performance verification?

8. What features in the WELL Building Standard are categorized as WELL organizational protocol?

WELL AP EXAM PRACTICE QUESTIONS

1. What is the level of certification achieved by a New and Existing Buildings project that has met all applicable preconditions and 60% of the applicable optimizations?

A. Silver
B. Gold
C. Platinum
D. None, the project fails the minimum requirements.

2. What is the second step in the timeline for the WELL certification process?

A. Register the project through WELL Online.
B. Assemble the project team for a charrette.
C. Submit the required documentation.
D. Have the WELL assessor visit the project site for performance verification.

3. The necessary documentation for a Core and Shell project has already been submitted, reviewed, and approved. What percentage of the project must now be occupied, at a minimum, before performance verification can be scheduled?

A. 0%, as it is unnecessary for Core and Shell performance verification
B. 50%
C. 75%
D. 100%, since the Core and Shell project type addresses the whole building

4. The WELL report for a New and Existing Buildings project indicates that the project has failed to meet the criteria for one of the applicable preconditions. After reviewing the report, the project owner determines that precondition had indeed been failed. What step can be taken to address the unmet criteria and reattempt the precondition?

A. Submit an appeal.
B. Accept the WELL report as is for a lower level of certification.
C. Initiate a curative action plan.
D. Pursue an AAP.

5. How many months does a project have to complete the recertification process after sending in the recertification application?

A. 3
B. 6
C. 9
D. 12

CHAPTER FOUR

AIR–PRECONDITIONS

Much of human health and wellness depends on air quality. However, outdoor air is becoming increasingly polluted worldwide due to activities such as fuel-burning transportation, construction, agriculture, and industry. The quality of indoor air can be affected by the pollution levels of outdoor air as well as indoor sources of pollutants and contaminants. The preconditions in the Air Concept of the WELL Building Standard establish minimum requirements for healthy indoor air quality by promoting clean air delivery through ventilation systems, minimizing the amount of pollutants and contaminants released into the indoor air of a building, and establishing effective cleaning and inspection policies.

OBJECTIVES

- Identify which features are preconditions in the Air Concept.
- Describe the health effects and required limitations of formaldehyde, volatile organic compounds (VOCs), particulate matter, carbon monoxide (CO), ozone (O_3), and radon for indoor air.
- Explain the health consequences of tobacco smoking and the strategies for prohibiting it on a project site.
- Describe the ventilation rates, filtration requirements, and other design strategies that ensure a ventilation system supplies a building with quality indoor air.
- Describe the strategies for preventing the presence of mold, microbes, and other pathogens in a building.
- Describe the strategies for managing or removing construction pollution, outdoor contaminants, and pesticides.
- Describe the health effects of lead, asbestos, polychlorinated biphenyls (PCBs), and mercury as well as abatement or removal practices.
- Identify the four ways moisture can enter a building and the strategies for managing the threat of water infiltration.

KEY TERMS

- allergen
- asbestos
- carbon monoxide (CO)
- coarse particle (PM_{10})
- fine particle ($PM_{2.5}$)
- formaldehyde
- high-touch surface
- lead
- mechanical ventilation
- mercury
- minimum efficiency reporting value (MERV)
- natural ventilation
- off-gassing
- ozone (O_3)
- particulate matter
- pathogen
- pesticide
- polychlorinated biphenyl (PCB)
- radon
- sick building syndrome (SBS)
- ultraviolet germicidal irradiation (UVGI)
- volatile organic compound (VOC)

Learner Resources

ATPeResources.com/QuickLinks
Access Code: 935534

AIR CONCEPT PRECONDITIONS

The air that people breathe has a significant impact on both their short- and long-term health. Improving and maintaining high air quality in an indoor environment is crucial to the health and wellness of building occupants. Therefore, WELL aims to promote clean air through the reduction and control of indoor air pollution and by providing the optimal level of indoor air quality.

The Air Concept is the largest concept in the WELL Building Standard. Many different design choices and policies can influence indoor air quality, which can then lead to health and wellness improvements for building occupants. The Air Concept contains 29 features that address design choices and policies. The first 12 features are predominantly preconditions, except for one optimization for New and Existing Interiors in Feature 08. **See Figure 4-1.**

Air is moved through building spaces by mechanical and natural means. Proper ventilation that provides clean, fresh air and removes old, stale air is essential for buildings such as commercial and institutional offices. Poor air quality in those types of buildings can impact productivity and lead to sick building syndrome. *Sick building syndrome (SBS)* is a set of symptoms, such as headaches, fatigue, eye irritation, and breathing difficulties, believed to be caused by indoor pollutants and poor environmental control that typically affects workers in modern airtight office buildings.

BY THE NUMBERS

- *A person breathes on average more than 15,000 liters of air every day.*

Mechanical ventilation is ventilation provided by mechanically powered equipment, such as motor-driven fans and blowers, but not by devices such as wind-driven turbine ventilators and mechanically operated windows. Mechanical ventilation systems use heating and cooling coils in conjunction with filters, fans, and ductwork to supply cooled or heated air to spaces. **See Figure 4-2.** Mechanical ventilation systems can be vectors for mold growth and can easily spread contaminants such as dust.

Natural ventilation is the movement of air into and out of a space primarily through intentionally provided openings (such as windows and doors), through nonpowered ventilators, or by infiltration. **See Figure 4-3.** Pressure differentials can be achieved through wind pressure or temperature differences. For example, a temperature pressure differential can be achieved because warm air rises since it is at a lower pressure than cold air. Natural ventilation systems are less common than mechanical ventilation systems because they are dependent on the quality and conditions of the outside air. Natural ventilation systems can easily allow the infiltration of humidity and air pollution, such as tobacco smoke and outdoor chemicals.

WELL Building Standard Features: Air Concept—Preconditions...						
	Project Type			Verification Documentation		
Features	Core and Shell	New and Existing Interiors	New and Existing Buildings	Letter of Assurance	Annotated Documents	On-Site Assessment
01, Air Quality Standards						
Part 1: Standards for Volatile Substances	P	P	P			Performance Test
Part 2: Standards for Particulate Matter and Inorganic Gases	P	P	P			Performance Test
Part 3: Radon	P	P	P			Performance Test

Figure 4-1. (*continued on next page*)

...WELL Building Standard Features: Air Concept—Preconditions...						
Features	Project Type			Verification Documentation		
	Core and Shell	New and Existing Interiors	New and Existing Buildings	Letter of Assurance	Annotated Documents	On-Site Assessment
02, Smoking Ban						
Part 1: Indoor Smoking Ban	P	P	P		Policy Document	
Part 2: Outdoor Smoking Ban	P	–	P			Visual Inspection
03, Ventilation Effectiveness						
Part 1: Ventilation Design	P	P	P	MEP		
Part 2: Demand Controlled Ventilation	P	P	P	MEP		
Part 3: System Balancing	–	P	P	MEP or Contractor		
04, VOC Reduction						
Part 1: Interior Paints and Coatings	P	P	P	Architect and Contractor		
Part 2: Interior Adhesives and Sealants	P	P	P	Architect and Contractor		
Part 3: Flooring	P	P	P	Architect and Contractor		
Part 4: Insulation	P	P	P	Architect and Contractor		
Part 5: Furniture and Furnishings	P	P	P	Architect and Owner		
05, Air Filtration						
Part 1: Filter Accommodation	P	P	P	MEP		Spot Check
Part 2: Particle Filtration	P	P	P	MEP		Spot Check
Part 3: Air Filtration Maintenance	P	P	P		Operations Schedule	
06, Microbe and Mold Control						
Part 1: Cooling Coil Mold Reduction	P	P	P		MEP Drawing or Operations Schedule	
Part 2: Mold Inspections	P	P	P			Visual Inspection
07, Construction Pollution Management						
Part 1: Duct Protection	P	P	P	Contractor		
Part 2: Filter Replacement	P	P	P	Contractor		
Part 3: Moisture Absorption Management	P	P	P	Contractor		
Part 4: Dust Containment and Removal	P	P	P	Contractor		
08, Healthy Entrance						
Part 1: Entryway Walk-Off Systems	P	O	P			Visual Inspection
Part 2: Entryway Air Seal	P	O	P			Visual Inspection
09, Cleaning Protocol						
Part 1: Cleaning Plan for Occupied Spaces	–	P	P		Operations Schedule	
10, Pesticide Management						
Part 1: Pesticide Use	P	–	P		Operations Schedule	

Figure 4-1. (*continued on next page*)

...WELL Building Standard Features: Air Concept—Preconditions						
Features	Project Type			Verification Documentation		
	Core and Shell	New and Existing Interiors	New and Existing Buildings	Letter of Assurance	Annotated Documents	On-Site Assessment
11, Fundamental Material Safety						
Part 1: Asbestos and Lead Restriction	P	P	P	Architect and MEP		
Part 2: Lead Abatement	P	P	P		Remediation Report	
Part 3: Asbestos Abatement	P	P	P		Remediation Report	
Part 4: Polychlorinated Biphenyl Abatement	P	P	P		Remediation Report	
Part 5: Mercury Limitation	P	P	P	MEP		
12, Moisture Management						
Part 1: Exterior Liquid Water Management	P	–	P		Professional Narrative	
Part 2: Interior Liquid Water Management	P	–	P		Professional Narrative	
Part 3: Condensation Management	P	–	P		Professional Narrative	
Part 4: Material Selection and Protection	P	–	P		Professional Narrative	

Figure 4-1. The Air Concept in WELL includes 12 preconditions that promote strategies such as setting standards for indoor air quality, increasing ventilation, reducing pollutants and toxic materials, and managing moisture infiltration to ensure building occupants have clean and safe indoor air to breathe.

MECHANICAL VENTILATION

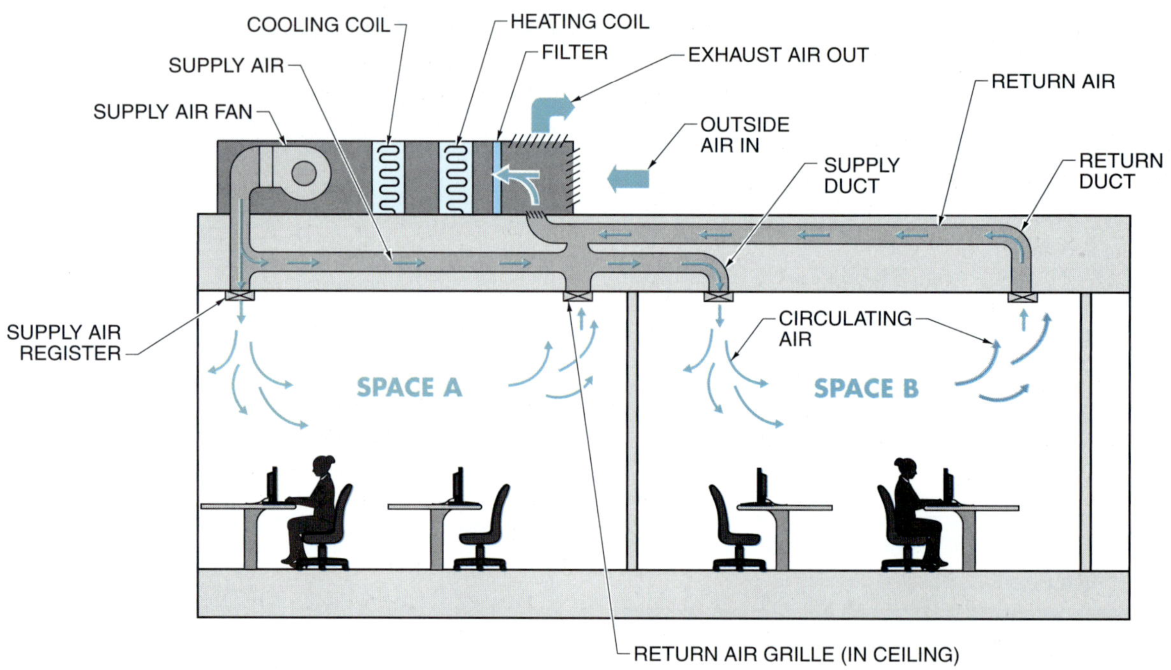

Figure 4-2. Mechanical ventilation is ventilation provided by mechanically powered equipment, such as motor-driven fans and blowers, but not by devices such as wind-driven turbine ventilators and mechanically operated windows.

NATURAL VENTILATION SYSTEMS

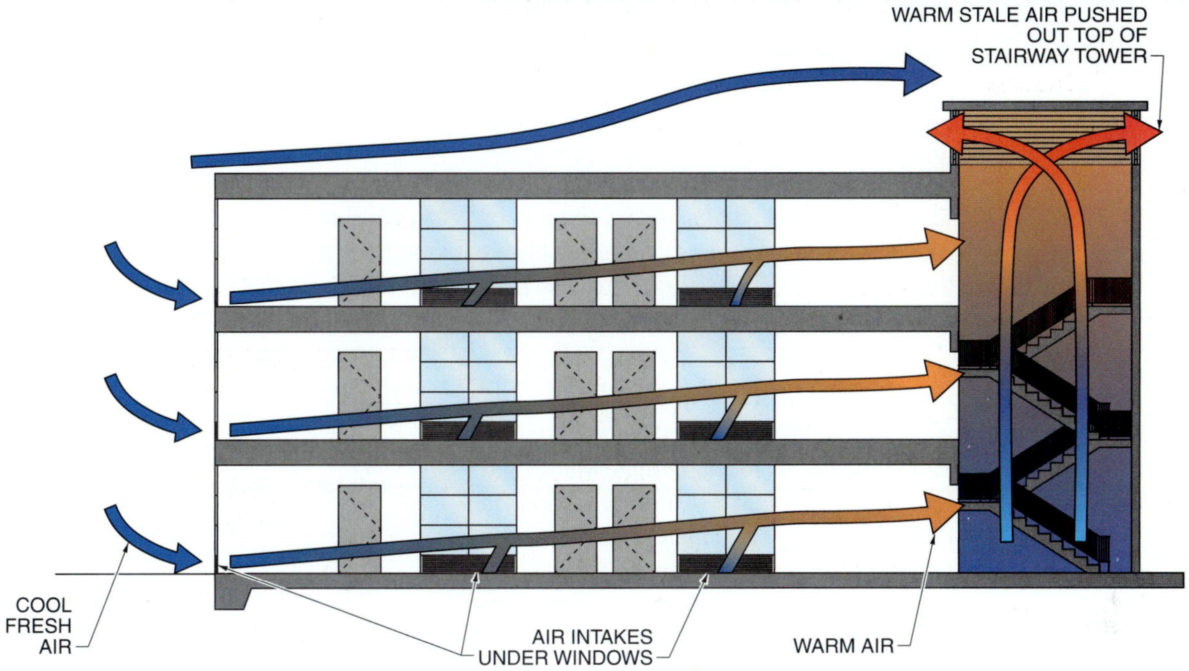

Figure 4-3. Natural ventilation systems use pressure differentials from temperature differences to move air through a building without the use of mechanical equipment.

FEATURE 01, AIR QUALITY STANDARDS

Feature 01, Air Quality Standards, is a precondition for all three project types that limits the levels of indoor air pollutants. Indoor pollutants can have immediate and long-lasting negative health effects. Exposure to excessive levels of air pollutants can cause short-term symptoms such as nausea, headaches, allergic reactions, and respiratory irritation. Long-term symptoms from exposure to air pollutants may include central nervous system damage, endocrine disruption, and cancer. Therefore, it is critical for a building to meet a basic level of higher indoor air quality.

BY THE NUMBERS

- *The levels of volatile organic compounds (VOCs) in an indoor environment can be five times higher than outdoor levels.*

Part 1: Standards for Volatile Substances

Part 1: Standards for Volatile Substances of Feature 01 addresses the levels of formaldehyde and volatile organic compounds (VOCs) in the air. Construction-related sources responsible for off-gassing these two pollutants include furniture materials, coatings, glues, and other products. *Off-gassing* is the release of chemicals or particulates into the air from substances and solvents used in the manufacture of a building product.

Formaldehyde is a colorless gas compound that is used for manufacturing melamine and phenolic resins, fertilizers, dyes, and embalming fluids. Short-term exposure to formaldehyde can cause eye, nose, or throat irritation, coughing or wheezing, and nausea. Chronic exposure to formaldehyde can cause severe skin irritation, respiratory problems, and an increased risk of cancer. Part 1a requires that a project limit the levels of formaldehyde in the air to less than 27 parts per billion (ppb).

A *volatile organic compound (VOC)* is a material containing carbon and hydrogen that evaporates and diffuses easily at ambient temperature and is emitted by a wide array of building materials, paints, wood preservatives, and other common consumer products. Exposure to VOCs can cause throat and nose irritation, headaches, nausea, liver damage, and an increased risk of cancer. Part 1b requires that a project limit the levels of VOCs in the air to less than 500 micrograms per cubic meter ($\mu g/m^3$).

Part 2: Standards for Particulate Matter and Inorganic Gases

Part 2: Standards for Particulate Matter and Inorganic Gases of Feature 01 addresses coarse and fine particulate matter and the inorganic gases of carbon monoxide and ozone. Each of these substances can be released into the air from various sources and can be dangerous. Part 2 sets limits for the levels of particulate matter and inorganic gases present in the air.

Carbon monoxide (CO) is a colorless, odorless, and highly poisonous gas formed by incomplete combustion. CO replaces oxygen in hemoglobin, which limits the ability of blood to deliver oxygen and can lead to death. CO sources include heating, ventilating, and air conditioning (HVAC) systems and water heaters that utilize combustion, gas engines, and fuel-burning appliances. Exposure to CO may cause headaches, fatigue, dizziness, drowsiness, or nausea. CO can be measured using a handheld meter. Part 2a requires that a project limit CO levels to less than 9 parts per million (ppm).

Particulate matter is a complex mixture of elemental and organic carbon, salts, mineral and metal dust, ammonia, and water that coagulate together into tiny solids and globules. A *coarse particle (PM_{10})* is particulate matter larger than 2.5 micrometers (µm) and smaller than 10 µm in diameter. A *fine particle ($PM_{2.5}$)* is particulate matter 2.5 µm or smaller in diameter. **See Figure 4-4.** These particles are inhalable and can cause respiratory distress, decreased lung function, irregular heartbeat, and an increased risk of lung cancer.

Particulate matter sources include dust, fly ash, soot, smoke, aerosols, or fumes. Particulate matter levels are measured using optical particle counters. Part 2b requires that a project limit $PM_{2.5}$ levels in the air to less than 15 $\mu g/m^3$. Part 2c requires that a project limit PM_{10} levels in the air to less than 50 $\mu g/m^3$.

Ozone (O_3) is the triatomic form of oxygen that is hazardous to the respiratory system at ground level. It can be released into the air by photocopiers, printers, and air cleaners. Short-term or low-level exposure to ozone can cause eye irritation, chest pain, coughing, shortness of breath, and throat irritation. Long-term or high-level exposure can cause lung tissue inflammation and the inability to fight respiratory infections. Building occupants with preexisting cardiorespiratory conditions such as asthma, bronchitis, or emphysema may be at increased risk of health complications even at low concentrations. The ozone level in the air can be measured using an indoor air quality monitor. Part 2d requires that a project limit the ozone level in the indoor air to less than 51 ppb.

WELLNESS FACT

To reduce the concentration of ozone in the air, office equipment such as photocopiers and printers should be placed in open, properly vented areas. Exhaust fans can also help protect occupants from exposure.

Part 3: Radon

Radon is a radioactive, carcinogenic noble gas generated from the decay of natural deposits of uranium. Radon is naturally emitted from soil or rocks and enters a building through cracks or openings in the floors or walls. Natural building materials such as granite tile or countertops, gypsum board, and concrete may emit minute levels of radon. Short-term radon exposure has no effects, but long-term, chronic exposure can lead to lung cancer. Radon levels can be measured using a radon gas detector or a test kit. Part 3 requires that a project limit the radon levels at its lowest occupied level to less than 0.148 Bq/L (4 pCi/L).

RELATIVE SIZES OF PARTICULATE MATTER

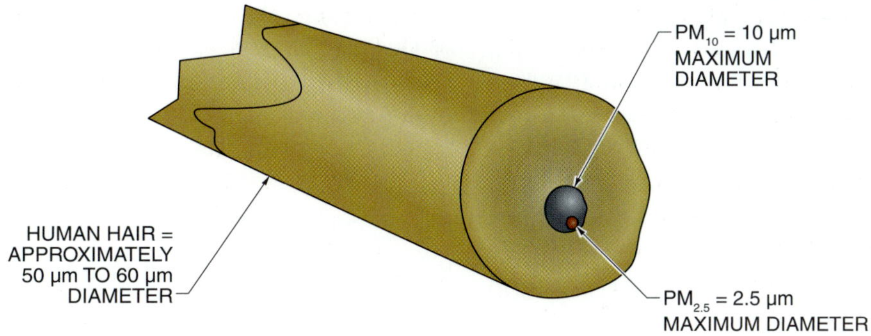

Figure 4-4. Particulate matter can vary in size and may or may not be able to be seen by the naked eye.

FEATURE 02, SMOKING BAN

Smoking tobacco products and secondhand inhalation of tobacco smoke can cause various diseases and negative health conditions. Because tobacco smoke directly enters the lungs, it presents a very high chance that the smoker will develop lung cancer after prolonged exposure. Other diseases and complications include chronic obstructive pulmonary disease (COPD), stroke, increased blood pressure, and other types of cancer.

The first part of Feature 02, Smoking Ban, is a precondition for all three project types, but the second part is only a precondition for Core and Shell and New and Existing Buildings project types. The goal of this feature is to eliminate smoking in a building and on the project site and to educate both occupants and visitors on the hazards of smoking and exposure to secondhand smoke.

Part 1: Indoor Smoking Ban

Part 1: Indoor Smoking Ban of Feature 02 bans all smoking and the use of e-cigarettes inside the project. E-cigarettes are small electronic devices that use a small heating element to vaporize nicotine-containing liquid. While the device does not burn any tobacco, the liquid and vapor still contain nicotine and other harmful chemicals.

Part 2: Outdoor Smoking Ban

Part 2: Outdoor Smoking Ban of Feature 02 requires the placement of signs that indicate a smoking ban near entrances and in specific exterior spaces. These signs must indicate that smoking is banned within 7.5 m (25′) of all entrances, operable windows, and building air intakes. **See Figure 4-5.** Signs must also indicate that smoking is banned in any regularly occupied exterior areas such as decks, patios, and accessible rooftops.

Part 2 also requires signs that educate occupants about the hazards of smoking in the specific areas where smoking is allowed (beyond the 7.5 m [25′] distance). These signs should be placed on all walkways within the area in intervals of 30 m (100′) at the most.

OUTDOOR SMOKING BAN

Figure 4-5. Feature 02, Smoking Ban, prohibits smoking indoors and limits outdoor smoking areas as indicated by signage.

BY THE NUMBERS

- *Even though smoking tobacco causes at least 40,000 premature deaths per year in the United States alone, at least 42 million U.S. adults (more than a billion worldwide) are smokers.*
- *A person who is a tobacco smoker has a life expectancy that is 10 years less than a nonsmoker.*
- *When cigarettes are smoked, their 600 ingredients form 7000 compounds that include at least 69 known carcinogens (cancer-causing substances).*

FEATURE 03, VENTILATION EFFECTIVENESS

Adequate ventilation is an important component of high indoor air quality and the health of building occupants. One of the key indicators of effective ventilation is the level of carbon dioxide (CO_2) in the spaces. The first and second parts of Feature 03, Ventilation Effectiveness, are preconditions for all three project types, while the third part is only a precondition for the New and Existing Interiors and New and Existing Buildings project types. This feature addresses the design of the ventilation system (HVAC system), maximum allowable CO_2 levels, and system testing and balancing.

Part 1: Ventilation Design

Part 1: Ventilation Design of Feature 03 uses ASHRAE Standard 62.1-2013, *Ventilation for Acceptable Indoor Air Quality*, for its requirements. Part 1a requires that the ventilation rates for mechanically ventilated projects meet the Ventilation Rate Procedure or IAQ Procedure within ASHRAE Standard 62.1-2013. Part 1b requires naturally ventilated projects to comply with the requirements for any procedure in ASHRAE Standard 62.1-2013.

Along with ASHRAE Standard 62.1-2013, Part 1b also uses the National Ambient Air Quality Standards (NAAQS) set forth by the U.S. Environmental Protection Agency (EPA) under the Clean Air Act for its requirements. Projects are required to demonstrate that the air surrounding the building for a distance of 1.6 km (1 mile) either complies with the NAAQS or passes the requirements of Feature 01, Air Quality Standards, for at least 95% of all hours in the previous year.

Part 2: Demand Controlled Ventilation

Part 2: Demand Controlled Ventilation of Feature 03 applies to any occupied spaces in the building that are 46.5 m^2 (500 ft^2) or larger and have an actual or expected occupant density of 25 people per 93 m^2 (1000 ft^2). Part 2 requires that projects limit CO_2 levels to below 800 ppm (measured at 1.2 m to 1.8 m [4′ to 6′] above the floor). Typically, outside air has CO_2 levels of 300 ppm to 400 ppm. Inadequate ventilation can cause excessive indoor CO_2 levels, which can lead to headaches, fatigue, or throat irritation. Part 2a applies to mechanical ventilation and Part 2b applies to natural ventilation.

Part 3: System Balancing

Part 3: System Balancing of Feature 03 requires that all HVAC systems (within the last five years) undergo or be scheduled to undergo testing and balancing after the substantial completion but before occupancy of a project. The testing and balancing process ensures that HVAC systems run at peak efficiency and meet the required ventilation requirements.

FEATURE 04, VOC REDUCTION

Feature 04, VOC Reduction, is a precondition for all three project types. The goal of this feature is to reduce the amount of materials, products, and finishes used in building construction that can release VOCs into the interior of the building. The five parts of this feature cover VOC reduction requirements for interior paints and coatings, interior adhesives and sealants, flooring, insulation, and furniture and furnishings. **See Figure 4-6.**

VOC Reduction Requirements		
Part	Amount	Applicable Standard
Part 1: Interior Paints and Coatings	**Meet one of the following for newly applied paints and coatings:**	
	100% of installed products	CARB 2007, SCM for Architectural Coatings or SCAQMD Rule 1113
	90% by volume	CDPH Standard Method v1.1-2010
	ALL	Applicable national VOC regulations or conduct testing of VOC content in accordance with ASTM D2369-10; ISO 11890, part 1; ASTM D6886-03; or ISO 11890-2
Part 2: Interior Adhesives and Sealants	**Meet one of the following for newly applied adhesives and sealants:**	
	100% of installed products	SCAQMD Rule 1168
	90% by volume	CDPH Standard Method v1.1-2010
	ALL	Applicable national VOC regulations or conduct testing of VOC content in accordance with ASTM D2369-10; ISO 11890, part 1; ASTM D6886-03; or ISO 11890-2
Part 3: Flooring	All newly installed flooring	CDPH Standard Method v1.1-2010
Part 4: Insulation	All newly installed thermal and acoustic insulation inside the waterproofing membrane	CDPH Standard Method v1.1-2010
Part 5: Furniture and Furnishings	95% by cost of newly purchased furniture and furnishings	ANSI/BIFMA e3-2011 Furniture Sustainability Standard, sections 7.6.1 and 7.6.2

Figure 4-6. Feature 04, VOC Reduction, requires that specific types of building products adhere to different standards regarding their VOC content.

Part 1: Interior Paints and Coatings

Part 1: Interior Paints and Coatings of Feature 04 requires that VOCs from newly applied interior paints and coatings be limited. Projects must meet one of the following requirements for paints and coatings:

- All (100%) of the installed products must either meet California Air Resources Board (CARB) 2007, Suggested Control Measure (SCM) for Architectural Coatings, or South Coast Air Quality Management District (SCAQMD) Rule 1113 for their VOC content.
- At a minimum, 90%, by volume, of the paints and coatings must meet California Department of Public Health (CDHP) Standard Method v1.1-2010 for their VOC emissions.
- The paints and coatings must meet applicable national VOC regulations or testing of VOC content must be performed in accordance with ASTM D2369-10; ISO 11890, part 1; ASTM D6886-03; or ISO 11890-2.

Part 2: Interior Adhesives and Sealants

Part 2: Interior Adhesives and Sealants of Feature 04 requires that VOCs from newly applied interior adhesives and sealants be limited. Although applied adhesives and sealants may not always be visible to building occupants, these products can release enough VOCs into the indoor air to become a health hazard. Projects must meet one of the following requirements for adhesives and sealants:

- All (100%) of the installed products must meet SCAQMD Rule 1168 for their VOC content.
- At a minimum, 90%, by volume, of the adhesives and sealants must meet CDHP Standard Method v1.1-2010 for their VOC emissions.
- The adhesives and sealants must meet applicable national VOC regulations or testing of VOC content must be performed in accordance with ASTM D2369-10; ISO 11890, part 1; ASTM D6886-03; or ISO 11890-2.

Part 3: Flooring

Part 3: Flooring of Feature 04 requires that the VOC content of newly installed flooring be limited. Projects must limit the VOC content of all newly installed interior floors in accordance with CDHP Standard Method v1.1-2010.

Part 4: Insulation

Part 4: Insulation of Feature 04 requires that the VOC content of newly installed interior thermal and acoustic insulation be limited. Projects must limit the VOC content of all newly installed insulation in accordance with CDHP Standard Method v1.1-2010.

Part 5: Furniture and Furnishings

Part 5: Furniture and Furnishings of Feature 04 requires that the VOC content of newly purchased interior furniture and furnishings within the scope of the project be limited. Projects must limit the VOC content of at least 95%, by cost, of all newly purchased furniture and furnishings per ANSI/BIFMA e3-2011 Furniture Sustainability Standard, sections 7.6.1 and 7.6.62, when tested in accordance with ANSI/BIFMA Standard Method M7.1-2011. Alternatively, the furniture and furnishings may meet CDPH Standard Method v1.1-2010.

FEATURE 05, AIR FILTRATION

Air filtration is an important component of a building's HVAC system. Air filters are designed to remove harmful contaminants from incoming outdoor and recirculated indoor air. Feature 05, Air Filtration, is a precondition for all three project types that addresses the minimum level of filtration, capability to upgrade filtration at a later date, and proper filter maintenance.

The two types of filters addressed in Feature 05 are carbon filters and media filters. Carbon filters use activated carbon to remove VOCs and other large particles. Media filters (also known as particle filters) use a fine polyester or fiberglass medium to remove small particles from the air and are classified by their minimum efficiency reporting value. *Minimum efficiency reporting value (MERV)* is a value assigned to an air filter that describes the amount of different types of particles removed when the filter is operating at the least effective point in its life. The higher the MERV number, the greater the air filter performance.

Part 1: Filter Accommodation

Part 1: Filter Accommodation of Feature 05 applies to projects in which recirculated air is used. These projects must have rack space available and identify rack locations for future implementation of carbon filters or combination particle/carbon filters in the main air ducts. Also, the mechanical system must be sized to accommodate additional filters.

Part 2: Particle Filtration

Part 2: Particle Filtration of Feature 05 requires that a project meet the requirements of either Part 2a or Part 2b. Part 2a requires the installation of MERV 13 filters (at a minimum) in HVAC systems for filtering outdoor air. **See Figure 4-7.** Part 2b requires that a project demonstrate that the levels of particulate matter in the air surrounding the project site for 1.6 km (1 mile) meet the standards set in Feature 01, Air Quality Standards, for at least 95% of all hours for the last calendar year.

Part 3: Air Filtration Maintenance

Part 3: Air Filtration Maintenance of Feature 05 requires that the project annually submit records of proper air filtration maintenance to the International WELL Building Institute (IWBI). Filters must be maintained as specified by the manufacturer's recommendations.

FEATURE 06, MICROBE AND MOLD CONTROL

Mold thrives in constantly wet environments. For example, HVAC systems contain cooling coils and drain pans that are constantly wet and present a natural breeding ground for molds and other organisms. If HVAC components are not properly maintained, mold may grow on them and release spores into the occupied spaces of a building. Mold spores can trigger asthma attacks, allergies, headaches, and other serious respiratory ailments. The presence of mold in an HVAC system may also decrease airflow and give the air a foul odor.

Mold may also be present elsewhere in the building. Areas that have continued exposure to moisture due to weather damage, leaks, or improper construction can also experience mold growth. In these cases, the mold may be difficult to locate and continue to build up behind walls or fixtures. Feature 06, Microbe and Mold Control, is a precondition for all three project types that addresses strategies to prevent mold growth and implement inspection, cleaning, and reporting procedures.

Part 1: Cooling Coil Mold Reduction

Part 1: Cooling Coil Mold Reduction of Feature 06 applies to buildings that use mechanical HVAC systems. Part 1a requires the use of ultraviolet germicidal irradiation devices, such as ultraviolet lamps, to control the growth of mold on cooling coils and drain pans. *Ultraviolet germicidal irradiation (UVGI)* is a sterilization method that uses ultraviolet (UV) light to break down microorganisms by destroying their DNA. **See Figure 4-8.** In order to prevent the generation of ozone, the wavelength of the UV lamps must be 254 nanometers (nm).

Part 1b requires that a building's HVAC system be inspected for mold growth on a quarterly basis. If mold growth is found, the HVAC system must be cleaned. Dated photographic evidence of the inspection must be submitted to IWBI every year for review.

Part 2: Mold Inspections

Part 2: Mold Inspections of Feature 06 requires that a mold inspection be performed by the WELL assessor. The WELL assessor must inspect for discoloration and mold on the ceilings, walls, and floors. The WELL assessor must also inspect for signs of water damage and pooling.

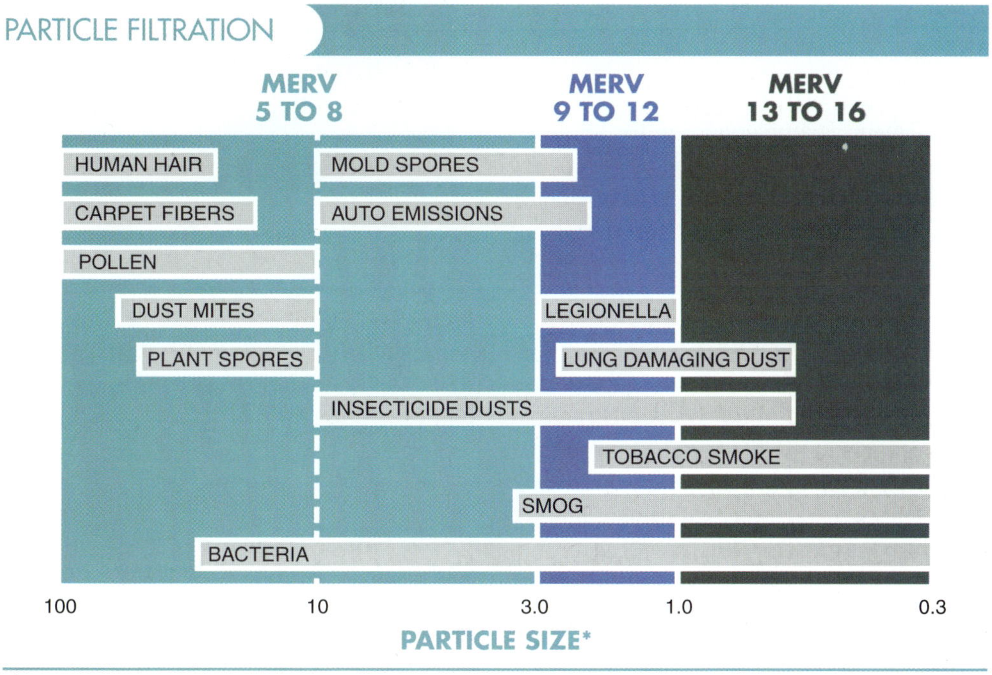

Figure 4-7. Part 2a of Feature 05, Air Filtration, requires that air filters installed in HVAC systems have a minimum MERV of 13 to filter out much of the particulate matter from outdoor air.

ULTRAVIOLET GERMICIDAL IRRADIATION (UVGI) DEVICES

Fresh-Aire UV

Figure 4-8. UVGI systems use UV light to break down microorganisms inside HVAC systems.

FEATURE 07, CONSTRUCTION POLLUTION MANAGEMENT

Construction activities produce large amounts of dust and dirt. This debris can be unintentionally spread throughout a building by its HVAC system. Also,water and moisture present during the construction process can be absorbed by some building materials and furnishings. Exposure to construction debris and problems from excess moisture can cause respiratory ailments and other health complications. Feature 07, Construction Pollution Management, is a precondition for all three project types that addresses the protection of ventilation system ductwork, use of filters, management of moisture absorption, and removal and containment of construction dust.

Part 1: Duct Protection

Part 1: Duct Protection of Feature 07 requires protective measures to prevent construction debris from entering the ventilation system. Per Part 1a, all newly installed ductwork must be sealed and protected from contamination while stored on site, as well as during and after installation, during construction. **See Figure 4-9.** If those proper preventive protection measures are not used, Part 1b requires that the ductwork be thoroughly cleaned before the registers, grilles, and diffusers are installed.

PROTECTED DUCTWORK

Figure 4-9. Part 1a of Feature 07, Construction Pollution Management, requires that ductwork be sealed and protected from contamination while being stored on site, as well as during and after installation.

Part 2: Filter Replacement

If a project's HVAC system runs during construction that occurs within one year before performance verification, Part 2: Filter Replacement of Feature 07 requires that the air filters be replaced with clean filters before occupancy. This ensures that contaminants do not enter the system. Clean air filters also help ensure that the system runs at peak efficiency.

Part 3: Moisture Absorption Management

Part 3 of Feature 07 addresses the need to prevent certain materials and products from absorbing water or moisture during the construction of a building that occurs within one year before performance verification. Porous, absorptive building materials and furnishings such as carpets, acoustical ceiling panels, wall coverings, and furniture may absorb water or moisture during the construction process. Part 3 requires that a separate area be designated to store and protect absorptive materials or products.

BY THE NUMBERS

- *Air pollution contributes to about 50,000 premature deaths in the United States and approximately 7 million premature deaths worldwide (or one in eight premature deaths), making it the largest global environmental health risk.*

Part 4: Dust Containment and Removal

Part 4: Dust Containment and Removal of Feature 07 addresses the containment or capture of dust and debris during construction that occurs within one year before performance verification. The first requirement is the physical separation of active construction spaces from other spaces through the sealing of openings or installation of temporary barriers. The second requirement is the installation of walk-off mats at entryways to reduce the dust, dirt, and other pollutants that may be spread on the soles of shoes or on the wheels of carts or dollies. The third requirement is that saws and other power tools must have guards or collectors to capture any generated dust or debris. **See Figure 4-10.**

FEATURE 08, HEALTHY ENTRANCE

Feature 08, Healthy Entrance, is a precondition for the Core and Shell and New and Existing Buildings project types. It is also an optimization for New and Existing Interiors. Little can be done to control the contaminants outside a building. However, strategies can be employed to limit the amount of outdoor contaminants that could enter the building either through the air or on the soles of occupants' shoes.

DUST CONTAINMENT

Festool USA

Figure 4-10. Part 4c of Feature 07, Construction and Pollution Management, requires that saws and other power tools have guards or collectors to capture any generated dust or debris.

Part 1: Entryway Walk-off Systems

Part 1: Entryway Walk-off Systems of Feature 08 requires that project teams implement one of three strategies for all regularly used building entrances. The chosen strategy must be maintained on a weekly basis. The strategy must also be as wide as the entrance width and at least 3 m (10′) long in the primary direction of travel (sum of indoor and outdoor length). The three strategies include the following:

- permanent entryway system comprising grilles, grates, or slots
- rollout mats
- material manufactured as an entryway walk-off system

Part 2: Entryway Air Seal

Part 2: Entryway Air Seal of Feature 8 requires that project teams implement one of three strategies to limit the movement of air from outdoors to indoors within mechanically ventilated main building entrances. The three strategies for air-sealing the entryways include the following:

- building entry vestibule with two normally closed doorways
- revolving entrance doors
- at least three normally shut doors that separate occupied spaces from the outdoors

FEATURE 09, CLEANING PROTOCOL

Maintaining a clean working environment after occupancy is an important part of WELL. Daily use by building occupants and visitors can leave debris and microscopic pathogens and allergens on the surface of spaces. A *pathogen* is an infectious biological agent such as a bacterium, virus, or fungus that is capable of causing disease in its host. An *allergen* is an environmental substance that can produce an allergic reaction in the body but may not be intrinsically harmful. Feature 09, Cleaning Protocol, is a precondition for the New and Existing Interiors and New and Existing Buildings project types that requires a cleaning plan to address which surfaces need to be cleaned and to what extent and frequency, as well as a list of approved cleaning products.

Part 1: Cleaning Plan for Occupied Spaces

Part 1: Cleaning Plan for Occupied Spaces of Feature 09 requires that a cleaning plan be created to reduce occupant exposure to pathogens, allergens, and harmful cleaning chemicals. The plan must contain the following elements:

- Part 1a requires the cleaning plan to include the Cleaning Equipment and Training section of Table A4 in Appendix C of the WELL Building Standard.
- Part 1b requires the plan to include a list of approved product seals with which all cleaning, disinfection, and hand-hygiene products must comply in accordance with the Cleaning, Disinfection and Hand Hygiene Product section in Table A4 in Appendix C.
- Part 1c requires the plan to include a list of high-touch surfaces and a schedule of sanitization or disinfection as specified in the Disinfection and Sanitization section in Table A4 in Appendix C. A *high-touch surface* is a surface that is frequently touched by building users and occupants. These surfaces include door handles, light switches, telephones, tabletops, and plumbing fixture handles.
- Part 1d requires the plan to include a cleaning schedule that specifies the extent and frequency of cleaning, including the Entryway Maintenance section of Table A4 in Appendix C.
- Part 1e requires dated, maintained cleaning logs to be available to all occupants.

FEATURE 10, PESTICIDE MANAGEMENT

Feature 10, Pesticide Management, is a precondition for the Core and Shell and New and Existing Buildings project types. This feature requires the creation of a pest management system that responsibly uses pesticides and prohibits the use of the most dangerous pesticides. A pesticide is a chemical used to destroy, repel, or control plants or animals. Short-term or low-level exposure to pesticides can cause headaches, dizziness, nausea, and irritation of the skin, nose, or throat. Long-term or high-level exposure can cause breathing difficulties, unconsciousness, loss of muscle control, kidney problems, and reproductive health issues or may lead to death.

BY THE NUMBERS

- *U.S. agricultural and commercial industries use about 1 billion pounds of pesticide every year.*

Part 1: Pesticide Use

Part 1: Pesticide Use of Feature 10 requires either the elimination of pesticide or herbicide use on outdoor plants or the minimization of hazards by following the requirements of Part 1a or Part 1b. Part 1a requires the creation of a pest management plan in place of pesticide or herbicide use. This plan should be based on Chapter 3 of the San Francisco Environment Code Integrated Pest Management (IPM) program. Part 1b requires that only pesticides with a hazard tier ranking of 3 (least hazardous) be used on the project site. This information can be found in Table A2 in Appendix C of the WELL Building Standard.

FEATURE 11, FUNDAMENTAL MATERIAL SAFETY

Feature 11, Fundamental Material Safety, is a precondition for all three project types. Hazardous materials such as lead, asbestos, polychlorinated biphenyls (PCBs), and mercury can be found in older building materials, fixtures, and equipment. The goal of this feature is to reduce or eliminate occupant exposure to these hazardous substances.

Lead is a naturally occurring metal found deep within the ground that was used in plumbing fixtures, lighting fixtures, and recycled building products. Although a product may be labeled as lead-free, it might still contain trace amounts of lead. Exposure to lead can cause damage to the central nervous system, cardiovascular system, reproductive system, hematological system, and kidneys. Lead exposure is especially detrimental to the cognitive development of children.

Asbestos is a naturally occurring mineral that was commonly used in insulation because of its chemical and flame resistance, tensile strength, and sound absorption properties. Asbestos can also be found in building materials such as fireproofing, gaskets, floor and ceiling tiles, and roofing felt. Long-term exposure to asbestos can increase the risk of mesothelioma, asbestosis, and lung cancer.

A *polychlorinated biphenyl (PCB)* is a former commercially produced synthetic organic chemical compound that may be present in products and materials produced before the 1979 PCB ban. PCBs can be found in electrical equipment and older fluorescent light fixtures. Exposure to PCBs may negatively affect the skin and impair liver functionality.

Mercury is a naturally occurring poisonous metal element that can be found in the earth's surface. Short-term mercury exposure can cause coughing, sore throat, chest pain, increased blood pressure or heart rate, and headaches. Long-term exposure can cause damage to the gastrointestinal tract, kidneys, and lungs. Mercury can also cause neurological and behavioral disorders. Similar to lead, mercury exposure can cause developmental and behavioral problems in children.

Part 1: Asbestos and Lead Restriction

Part 1: Asbestos and Lead Restriction of Feature 11 addresses the levels of lead and asbestos in new building materials installed in the project. Part 1a requires that there be no asbestos-containing materials installed in the project. Part 1b limits the amount of lead in components for a plumbing system that provides water intended for human consumption. Lead in the wetted surfaces of pipes, fittings, and fixtures is limited to a weighted average of 0.25%. Solder or flux is limited to 0.20%. Part 1c limits the amount of lead in all other building materials to no more than 100 ppm (by weight).

Part 2: Lead Abatement

Part 2: Lead Abatement of Feature 11 addresses the guidelines for the identification and abatement of lead in projects. Part 2 applies to any repair, renovation, demolition, or painting of projects that were constructed prior to any laws banning or restricting lead paint. The standards used for lead abatement in WELL are the U.S. EPA lead standards, EPA 40 CFR Part 745.65 and EPA 40 CFR Part 745.277. These two standards apply to the following activities:

- EPA 40 CFR Part 745.65 is used as the basis for on-site investigations of lead in buildings.
- EPA 40 CFR Part 745.227 is used as the basis for work practices and lead abatement activities.

Note: Any future rulings from the EPA regarding lead repair, renovation, and painting supersedes these standards. Part 2 requires adherence to any final rules.

Part 3: Asbestos Abatement

Part 3: Asbestos Abatement of Feature 11 contains guidelines for asbestos abatement in projects that were constructed prior to applicable laws concerning asbestos. Part 3a requires that an asbestos inspection be performed every three years. An accredited asbestos professional must perform this inspection per one of the following:

- Asbestos Hazard Emergency Response Act's (AHERA's) Asbestos Model Accreditation Plan (MAP)
- National Emissions Standards for Hazardous Air Pollutants (NESHAP)
- accredited asbestos consultant (state or local equivalent)

Part 3b requires the development, maintenance, and update of asbestos management plans in accordance with AHERA. The asbestos management plans must address minimizing asbestos hazards through abatement procedures outlined in Asbestos-Containing Materials in Schools (EPA 40 CFR Part 763). Part 3c requires that post-abatement clearance be conducted in accordance with AHERA and Asbestos-Containing Materials in Schools (EPA 40 CFR Part 763).

Part 4: Polychlorinated Biphenyl Abatement

Part 4: Polychlorinated Biphenyl Abatement of Feature 11 contains guidelines for the identification and removal of PCBs in projects that were built or renovated between 1950 and the time any applicable laws limiting PCBs were enacted. Part 4a requires the evaluation and abatement of materials in accordance with the U.S. EPA's Steps to Safe PCB Abatement Activities. Part 4b requires the removal and safe disposal of PCB-containing fluorescent light ballasts in accordance with the EPA's guidelines.

Part 5: Mercury Limitation

Part 5: Mercury Limitation of Feature 11 sets restrictions for the use of mercury-containing materials. **See Figure 4-11.** A project must adhere to the following restrictions:

- Thermometers, switches, or electrical relays that contain mercury are not specified or used.
- Lamps are installed that comply with the low-mercury limits from Table A5 in Appendix C of the WELL Building Standard. A plan must be developed to replace any noncompliant lamps with low-mercury or mercury-free lamps.

- Only illuminated exit signs that use light-emitting diode (LED) or light-emitting capacitor (LEC) lamps are specified.
- Mercury vapor or probe-start, metal halide high-intensity discharge (HID) lamps are not used.

MERCURY LIMITATION

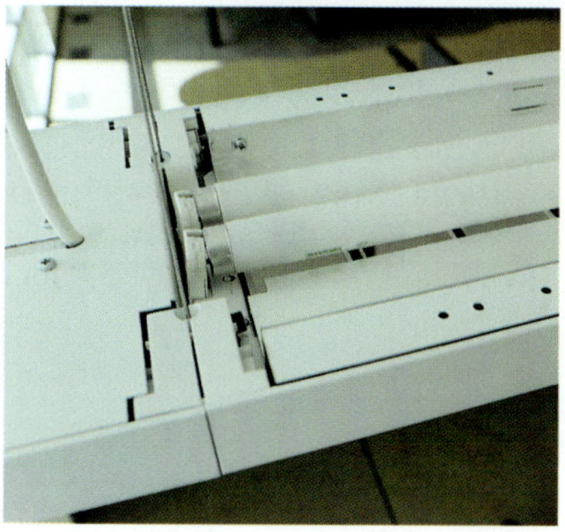

Figure 4-11. One requirement for limiting possible exposure to mercury vapor is to replace mercury-containing lamps with low-mercury or mercury-free fluorescent lamps.

FEATURE 12, MOISTURE MANAGEMENT

Feature 12, Moisture Management, is a precondition for the Core and Shell and New and Existing Building project types that addresses moisture management through the principles found in the U.S. EPA's Moisture Control Guidance for Building Design, Construction and Maintenance. Moisture infiltration into the interior of a building can cause structural damage, mold growth, and negative effects on indoor air quality. **See Figure 4-12.** Signs of moisture infiltration include stained foundation walls, interior walls, or ceilings; bubbling or peeling drywall or plaster; visible mold or mildew; or a musty smell. The four ways that moisture can infiltrate a building include the following:

- **Bulk water.** Bulk water enters a building in the form of leaks. These leaks can be due to a damaged, unmaintained, or incorrectly specified building envelope. Bulk water can also enter the building through plumbing or malfunctioning HVAC systems.
- **Capillary water.** Capillary water enters a building through porous building materials and finishes. Building materials such as concrete, mortar, and insulation can allow moisture infiltration via capillary action.
- **Air-transported moisture.** Air-transported moisture is commonly known as water vapor or humidity. When indoor air is heavy with water vapor, cool surfaces draw in the vapor and cause it to condense onto the surfaces. As this occurs, the condensed water may drip onto other surfaces or materials.
- **Vapor diffusion.** Vapor diffusion is the infiltration of water vapor through a vapor-permeable material. Materials such as gypsum board, concrete blocks, and brick are examples of construction materials that allow vapor diffusion.

MOISTURE INFILTRATION

Figure 4-12. Moisture infiltration into the interior of a building can cause structural damage, mold growth, and negative effects on indoor air quality.

Part 1: Exterior Liquid Water Management

Part 1: Exterior Liquid Water Management of Feature 12 requires that a project team develop a narrative addressing the infiltration of water from exterior sources. The narrative should include the location of the project site and the climate and weather for the area. This narrative must address the following four concerns:

- site drainage and irrigation
- local water table
- building penetrations, such as windows, and mechanical, electrical, and plumbing (MEP) penetrations
- porous building materials connected to exterior sources of water

Part 2: Interior Liquid Water Management

Part 2: Interior Liquid Water Management of Feature 12 addresses the management of interior sources of water. Part 2 requires that the project team develop another narrative describing how to manage possible sources of interior water. This narrative must address the following four concerns:

- plumbing leaks
- appliances, such as clothes washers, directly connected to the water supply
- porous building materials connected to interior sources of water
- new building materials that have high "built-in" moisture content or have been wetted during construction and brought into the interior

Part 3: Condensation Management

Part 3: Condensation Management of Feature 12 also requires the project team to develop a narrative addressing sources of condensation in the building. This narrative must address the following four concerns:

- high interior relative humidity levels
- air leakage that could wet exposed interior materials or hidden interstitial materials through condensation
- cooler surfaces such as basement or slab-on-grade floors or closets/cabinets on exterior walls
- oversized AC units that cycle on and off too quickly, which prevents moisture from being removed from the interior air

Part 4: Material Selection and Protection

Part 4: Material Selection and Protection of Feature 12 requires that the project team develop a narrative describing the use of moisture-tolerant materials and the protection of moisture-sensitive materials. The narrative must address the following five concerns:

- exposed entryways and glazing
- porous cladding materials
- finished floors in potentially damp or wet rooms
- interior sheathing in potentially damp or wet rooms
- sealing and storing of absorptive materials during construction

KEY TERMS AND DEFINITIONS

allergen: An environmental substance that can produce an allergic reaction in the body but may not be intrinsically harmful.

asbestos: A naturally occurring mineral that was commonly used in insulation because of its chemical and flame resistance, tensile strength, and sound absorption properties.

carbon monoxide (CO): A colorless, odorless, and highly poisonous gas formed by incomplete combustion.

coarse particle (PM_{10}): Particulate matter larger than 2.5 micrometers (μm) and smaller than 10 μm in diameter.

fine particle ($PM_{2.5}$): Particulate matter 2.5 micrometers (μm) or smaller in diameter.

formaldehyde: A colorless gas compound that is used for manufacturing melamine and phenolic resins, fertilizers, dyes, and embalming fluids.

high-touch surface: A surface that is frequently touched by building users and occupants, including door handles, light switches, telephones, tabletops, and plumbing fixture handles.

lead: A naturally occurring metal found deep within the ground that was used in plumbing fixtures, lighting fixtures, and recycled building products.

mechanical ventilation: A ventilation provided by mechanically powered equipment, such as motor-driven fans and blowers, but not by devices such as wind-driven turbine ventilators and mechanically operated windows.

mercury: A naturally occurring poisonous metal element that can be found in the earth's surface.

minimum efficiency reporting value (MERV): A value assigned to an air filter that describes the amount of different types of particles removed when the filter is operating at the least effective point in its life.

natural ventilation: The movement of air into and out of a space primarily through intentionally provided openings (such as windows and doors), through nonpowered ventilators, or by infiltration.

off-gassing: The release of chemicals or particulates into the air from substances and solvents used in the manufacture of a building product.

ozone (O_3): The triatomic form of oxygen that is hazardous to the respiratory system at ground level.

particulate matter: A complex mixture of elemental and organic carbon, salts, mineral and metal dust, ammonia, and water that coagulate together into tiny solids and globules.

pathogen: An infectious biological agent such as a bacterium, virus, or fungus that is capable of causing disease in its host.

pesticide: A chemical used to destroy, repel, or control plants or animals.

polychlorinated biphenyl (PCB): A former commercially produced synthetic organic chemical compound that may be present in products and materials produced before the 1979 PCB ban.

radon: A radioactive, carcinogenic noble gas generated from the decay of natural deposits of uranium.

sick building syndrome (SBS): A set of symptoms, such as headaches, fatigue, eye irritation, and breathing difficulties, believed to be caused by indoor pollutants and poor environmental control that typically affects workers in modern airtight office buildings.

ultraviolet germicidal irradiation (UVGI): A sterilization method that uses ultraviolet (UV) light to break down microorganisms by destroying their DNA.

volatile organic compound (VOC): A material containing carbon and hydrogen that evaporates and diffuses easily at ambient temperature and is emitted by a wide array of building materials, paints, wood preservatives, and other common consumer products.

CHAPTER REVIEW

Completion

______________________ 1. The first ___ features of the Air Concept are preconditions.

______________________ 2. Part 2: Outdoor Smoking Ban of Feature 02, Smoking Ban, requires the placement of signs that indicate a smoking ban in specific exterior spaces and within ___ of all entrances, operable windows, and building air intakes.

______________________ 3. Part 1: Ventilation Design of Feature 03, Ventilation Effectiveness, uses ASHRAE Standard ___, *Ventilation for Acceptable Indoor Air Quality*, for its requirements.

______________________ 4. Part 2: Demand Controlled Ventilation of Feature 03, Ventilation Effectiveness, applies to any occupied spaces in the building that are 46.5 m^2 (500 ft^2) or larger and have an actual or expected occupant density of ___ people per 93 m^2 (1000 ft^2).

______________________ 5. ___ is a value assigned to an air filter to describe the amount of different types of particles removed when the filter is operating at the least effective point in its life.

______________________ 6. Part 3: Air Filtration Maintenance of Feature 05, Air Filtration, requires that records of proper air filtration maintenance be submitted annually to the ___.

______________________ 7. If HVAC components are not properly maintained, ___ may grow on them and release spores into the occupied spaces of a building, which can trigger asthma attacks, allergies, headaches, and other serious respiratory ailments.

______________________ 8. Feature 07, ___, is a precondition for all three project types that addresses the protection of ventilation system ductwork, use of filters, management of moisture absorption, and removal and containment of construction dust.

______________________ 9. The strategy that a project team chooses to implement to meet the requirements of Part 1: Entryway Walk-off Systems of Feature 08, Healthy Entrance, must be as wide as the entrance and at least ___ long in the primary direction of travel (sum of indoor and outdoor length).

______________________ 10. A(n) ___ is a surface that is frequently touched by building users and occupants, including door handles, light switches, telephones, tabletops, and plumbing fixture handles.

______________________ 11. Part 1: Pesticide Use of Feature 10, Pesticide Management, requires either the elimination of pesticide or herbicide use on outdoor plants or the minimization of hazards. Hazards can be minimized by either creating a pest management plan based on Chapter 3 of the San Francisco Environment Code Integrated Pest Management (IPM) program or by using only pesticides with a hazard tier ranking of ___ on the project site.

______________________ 12. Part 1: Asbestos and Lead Restriction of Feature 11, Fundamental Material Safety, requires that there be no asbestos and no more than ___ ppm (by weight) of lead in any new building materials installed in a project.

______________________ 13. Part 3: Asbestos Abatement of Feature 11, Fundamental Material Safety, requires that an asbestos inspection be performed by an accredited professional every ___ years.

______________________ 14. Part 4: Polychlorinated Biphenyl Abatement of Feature 11, Fundamental Material Safety, requires the removal and safe disposal of PCB-containing ___ light ballasts in accordance with the EPA's guidelines.

CHAPTER REVIEW *(continued)*

Short Answer

1. List the limits placed on the levels of formaldehyde, volatile organic compounds (VOCs), carbon monoxide (CO), fine particles ($PM_{2.5}$), coarse particles (PM_{10}), ozone (O_3), and radon that can be in the indoor air of a building.

2. List each of the five parts for Feature 04, VOC Reduction.

3. Where are the different locations that must be inspected for mold in a mechanically ventilated building and what are the signs of water infiltration?

4. List the strategies that fulfill the requirements of Part 1: Entryway Walk-off Systems and Part 2: Entryway Air Seal of Feature 08, Healthy Entrance.

5. What are the elements that must be included in a cleaning plan per Part 1: Cleaning Plan for Occupied Spaces of Feature 09, Cleaning Protocol?

6. List the U.S. EPA lead standards and for which activities they are used to fulfill the requirements of Part 2: Lead Abatement of Feature 11, Fundamental Material Safety.

7. What are the restrictions a project must adhere to per Part 5: Mercury Limitation of Feature 11, Fundamental Material Safety?

8. List the four concerns that must be addressed by the narrative developed for Part 1: Exterior Liquid Water Management of Feature 12, Moisture Management.

CHAPTER REVIEW *(continued)*

9. List the four concerns that must be addressed by the narrative developed for Part 2: Interior Liquid Water Management of Feature 12, Moisture Management.

10. List the four concerns that must be addressed by the narrative developed for Part 3: Condensation Management of Feature 12, Moisture Management.

11. List the five concerns that must be addressed by the narrative developed for Part 4: Material Selection and Protection of Feature 12, Moisture Management.

WELL AP EXAM PRACTICE QUESTIONS

1. A project has complied with all of the requirements for procedures in ASHRAE Standard 62.1-2013, and it is currently being determined whether the ambient air around the project complies with the U.S. EPA's National Ambient Air Quality Standards (NAAQS). Within what distance of the project must the ambient air quality comply with the NAAQS to meet Part 1: Ventilation Design of Feature 03, Ventilation Effectiveness?
   - A. 0.4 km (¼ mile)
   - B. 0.8 km (½ mile)
   - C. 1.2 km (¾ mile)
   - D. 1.6 km (1 mile)

2. What is the minimum MERV rating an air filter must have to fulfill the first requirement of Part 1: Filter Accommodation of Feature 05, Air Filtration?
   - A. MERV 8
   - B. MERV 11
   - C. MERV 13
   - D. MERV 15

3. Which feature from the Air Concept is a precondition that includes a part that must be verified by a mechanical, electrical, and plumbing (MEP) drawing or an operations schedule?
   - A. Feature 02, Smoking Ban, Part 1: Indoor Smoking Ban
   - B. Feature 05, Air Filtration, Part 3: Air Filtration Maintenance
   - C. Feature 06, Microbe and Mold Control, Part 1: Cooling Coil Mold Reduction
   - D. Feature 09, Cleaning Protocol, Part 1: Cleaning Plan for Occupied Spaces

4. The ductwork for a project was stored on site during construction activities. If the ducts were not sealed or protected from contamination, what must be done in order to meet the requirements of Part 1: Duct Protection for Feature 07, Construction Pollution Management?
   - A. MERV 8 filters must be installed to filter out the large, construction-generated particulate matter.
   - B. The ductwork must be vacuumed out before registers, grilles, and diffusers are installed.
   - C. An ultraviolet (UV) light must be used to sterilize the ductwork.
   - D. Nothing, the project team has failed to meet the requirements of this feature.

5. Which feature from the Air Concept is a precondition that restricts the presence of lead, asbestos, PCBs, and mercury in a project and in the building materials?
   - A. Feature 01, Air Quality Standards
   - B. Feature 04, VOC Reduction
   - C. Feature 07, Construction Pollution Management
   - D. Feature 11, Fundamental Material Safety

CHAPTER FIVE

AIR–OPTIMIZATIONS

Projects that achieve the preconditions of the Air Concept can further improve indoor air quality by implementing the strategies found in the optimizations of the Air Concept. Many different building products, materials, and policies can contribute to better indoor air and human health and wellness. The goal of the optimizations in the Air Concept of the WELL Building Standard is to promote strategies for increasing the supply of fresh air, controlling humidity, implementing air monitors, purifying air, and reducing the exposure of building occupants to toxic materials and harmful substances.

OBJECTIVES

- Identify which features are optimizations in the Air Concept.
- Explain the importance of air flushing.
- Explain the consequences of humidity and air leakage and how they can be controlled or managed.
- Describe the strategies that can be used to measure and maintain air quality and building performance.
- Describe the importance of operable windows and the strategies that can be employed to maintain indoor air quality when they are designed into a building.
- Describe alternative ventilation methods such as dedicated outdoor air systems (DOASs) and displacement ventilation systems.
- Explain the nonchemical ways to control pests in a building.
- Identify the strategies that can be employed to purify and sanitize the air in a building space.
- Explain how toxic pollutants from combustion, building products, and materials can be managed or reduced.
- Describe the best practices and strategies for achieving a clean indoor environment.

KEY TERMS

- air exfiltration
- air flushing
- air infiltration
- dedicated outdoor air system (DOAS)
- displacement ventilation system
- halogenated flame retardant
- high-efficiency particulate air (HEPA) filter
- isocyanate
- perfluorinated compounds (PFCs)
- photocatalytic oxidation system
- phthalate
- polybrominated diphenyl ethers (PBDEs)
- polyurethane
- urea-formaldehyde

Learner Resources

ATPeResources.com/QuickLinks
Access Code: 935534

AIR CONCEPT OPTIMIZATIONS

The Air Concept is the largest concept in the WELL Building Standard with 29 features that address design choices and policies to protect and improve indoor air quality. The first 12 features of the Air Concept contain the preconditions that are applicable to the three project types as well as one optimization in Feature 08, Healthy Entrance for the New and Existing Interior project type. Feature 13 through Feature 29 include all of the other optimizations for the Air Concept. **See Figure 5-1.** These optimizations use the designs and policies implemented through the 12 preconditions to provide further protections and improvements for indoor air quality. The optimizations address topics such as indoor quality, filtration, ventilation, toxic materials, and cleaning.

After the construction of a building is complete but before occupancy, indoor air quality can greatly benefit from an air flush using the building's ventilation system.

FEATURE 08, HEALTHY ENTRANCE

Feature 08, Healthy Entrance, is only an optimization for the New and Existing Interiors project type. Otherwise, it is a precondition for the Core and Shell and New and Existing Buildings project types. Exam candidates should read Chapter 4 in this exam preparation guide for detailed information on the Air Concept preconditions and this one optimization.

FEATURE 13, AIR FLUSH

Feature 13, Air Flush, is an optimization for all three project types. *Air flushing* is a technique used to remove or reduce airborne contaminants and pollutants by running the ventilation system for an extended period of time after construction is complete but before occupancy. The continuous operation of the ventilation system removes volatile organic compounds (VOCs) and other pollutants that may have become airborne due to construction-related activities or the off-gassing of materials or products.

Part 1: Air Flush

Part 1: Air Flush of Feature 13 sets the requirements for the air flush process. For the air flush, the interior temperature and humidity of the building must be maintained at a minimum of 15°C (59°F) and no more than 60% relative humidity. Project teams can decide whether to complete the air flush entirely prior to occupancy or to split the process between before occupancy and after occupancy begins. **See Figure 5-2.**

Part 1a is the option to perform the air flush entirely prior to occupancy. However, construction activities must be complete before the air flush can be performed. The air flush for Part 1a requires a total volume of outdoor air of 4266 m^3 per m^2 of floor area (14,000 ft^3 per ft^2).

Part 1b requires a two-part air flush—the first part before occupancy and the second part after the building has been occupied. The first part of the air flush requires an outdoor air volume of 1066 m^3 per m^2 of floor area (3500 ft^3 per ft^2). The second part of the air flush requires 3200 m^3 per m^2 of floor area (10,500 ft^3 per ft^2). During the second part, the ventilation system must provide at least 0.1 m^3 per minute of outdoor air per m^2 of floor area (0.3 cfm outdoor air per ft^2) at all times.

WELL Building Standard Features: Air Concept—Optimizations...						
Features	**Project Type**			**Verification Documentation**		
	Core and Shell	**New and Existing Interiors**	**New and Existing Buildings**	**Letter of Assurance**	**Annotated Documents**	**On-Site Assessment**
13, Air Flush						
Part 1: Air Flush	–	O	O	Contractor		
14, Air Infiltration Management						
Part 1: Air Leakage Testing	O	O	O		Commissioning Report	
15, Increased Ventilation						
Part 1: Increased Outdoor Air Supply	O	O	O	MEP		
16, Humidity Control						
Part 1: Relative Humidity	–	O	O	MEP		Spot Measurement
17, Direct Source Ventilation						
Part 1: Pollution Isolation and Exhaust	O	O	O		Mechanical and Architectural Drawings	
18, Air Quality Monitoring and Feedback						
Part 1: Indoor Air Monitoring	–	O	O	MEP		Spot Check
Part 2: Air Data Record Keeping and Response	–	O	O		Operations Schedule	
Part 3: Environmental Measures Display	–	O	O			Visual Inspection
19, Operable Windows						
Part 1: Full Control	O	O	O	Architect		Spot Check
Part 2: Outdoor Air Measurement	O	O	O		Policy Document	
Part 3: Window Operation Management	O	O	O	Owner		Spot Check
20, Outdoor Air Systems						
Part 1: Dedicated Outdoor Air Systems	O	O	O	MEP		
21, Displacement Ventilation						
Part 1: Displacement Ventilation Design and Application	–	O	O	MEP		
Part 2: System Performance	–	O	O	MEP		
22, Pest Control						
Part 1: Pest Reduction	–	O	O		Operations Schedule	Spot Check
Part 2: Pest Inspection	–	O	O			Visual Inspection
23, Advanced Air Purification						
Part 1: Carbon Filtration	O	O	O	MEP		Spot Check
Part 2: Air Sanitization	O	O	O	MEP		Spot Check
Part 3: Air Quality Maintenance	O	O	O		Operations Schedule	
24, Combustion Minimization						
Part 1: Appliance and Heater Combustion Ban	O	O	O	Owner		Spot Check
Part 2: Low-Emission Combustion Sources	O	–	O	MEP		
Part 3: Engine Exhaust Reduction	O	–	O			Visual Inspection
Part 4: Construction Equipment	O	–	O	Contractor		

Figure 5-1. (*continued on next page*)

...WELL Building Standard Features: Air Concept—Optimizations						
Features	Project Type			Verification Documentation		
	Core and Shell	New and Existing Interiors	New and Existing Buildings	Letter of Assurance	Annotated Documents	On-Site Assessment
25, Toxic Material Reduction						
Part 1: Perfluorinated Compound Limitation	–	O	O	Architect and Owner		
Part 2: Flame Retardant Limitation	–	O	O	Architect and Contractor		
Part 3: Phthalate (Plasticizers) Limitation	–	O	O	Architect and Contractor		
Part 4: Isocyanate-Based Polyurethane Limitation	–	O	O	Architect and Contractor		
Part 5: Urea-Formaldehyde Restriction	–	O	O	Architect and Contractor		
26, Enhanced Material Safety						
Part 1: Precautionary Material Selection	–	O	O	Architect, Contractor, and Owner		
27, Antimicrobial Activity for Surfaces						
Part 1: High-Touch Surfaces	–	O	O		Architectural Drawing or Operations Schedule	
28, Cleanable Environment						
Part 1: Material Properties	–	O	O	Architect and Contractor		Spot Check
Part 2: Cleanability	–	O	O	Architect and Contractor		Spot Check
29, Cleaning Equipment						
Part 1: Equipment and Cleaning Agents	–	O	O		Operations Schedule	
Part 2: Chemical Storage	–	O	O			Visual Inspection

Figure 5-1. Of the Air Concept's 29 features addressing design choices and policies for protected and improved indoor air quality, Feature 13 through Feature 29 include most of the optimizations. Feature 08, Healthy Entrance, is only an optimization for the New and Existing Interiors project type.

FEATURE 14, AIR INFILTRATION MANAGEMENT

Feature 14, Air Infiltration Management, is an optimization for all three project types that requires building envelope commissioning and a plan for action and remediation. Air infiltration can reduce the indoor air quality of the building and increase the cost of heating or cooling. *Air infiltration* is the movement of air into a conditioned building space through an unwanted void in the building envelope. *Air exfiltration* is the movement of air out of a conditioned building space.

In warm weather, air infiltration can bring humid outdoor air into a building. In cold weather, warm, moist air can leak (exfiltrate) from the cold building envelope. Both of these situations can result in condensation that can lead to material damage or the development of mold. In addition, outside air infiltrating occupied spaces may carry pollutants or other contaminants that can affect the occupants.

Part 1: Air Leakage Testing

Part 1: Air Leakage Testing of Feature 14 requires that both building envelope commissioning and the plan for action and remediation be performed after substantial completion of the building but before occupancy. In addition, building envelope commissioning must be performed in accordance with ASHRAE Guideline 0-2005 and the National Institute of Building Sciences (NIBS) Guideline 3-2012. The plan for action and remediation should address what should be done if unacceptable conditions are found during the commissioning process.

FEATURE 15, INCREASED VENTILATION

Feature 15, Increased Ventilation, is an optimization for all three project types. Although projects that meet Feature 03, Ventilation Effectiveness, a precondition for the Air Concept, have already achieved acceptable ventilation levels, this optimization provides the opportunity for increased outdoor air ventilation. These increases in the supply of outdoor air will help to provide for optimal indoor air quality.

Part 1: Increased Outdoor Air Supply

Part 1: Increased Outdoor Air Supply of Feature 15 applies to the rate of the outdoor air supply for all regularly occupied spaces. One of the following options is required to meet this feature. Part 1a requires that the ventilation supply rates exceed the rates met in Part 1a of Feature 03, Ventilation Effectiveness, by 30%. Part 1b requires the project to follow CIBSE AM10, Section 4, to predict that room-by-room airflows will provide effective natural ventilation.

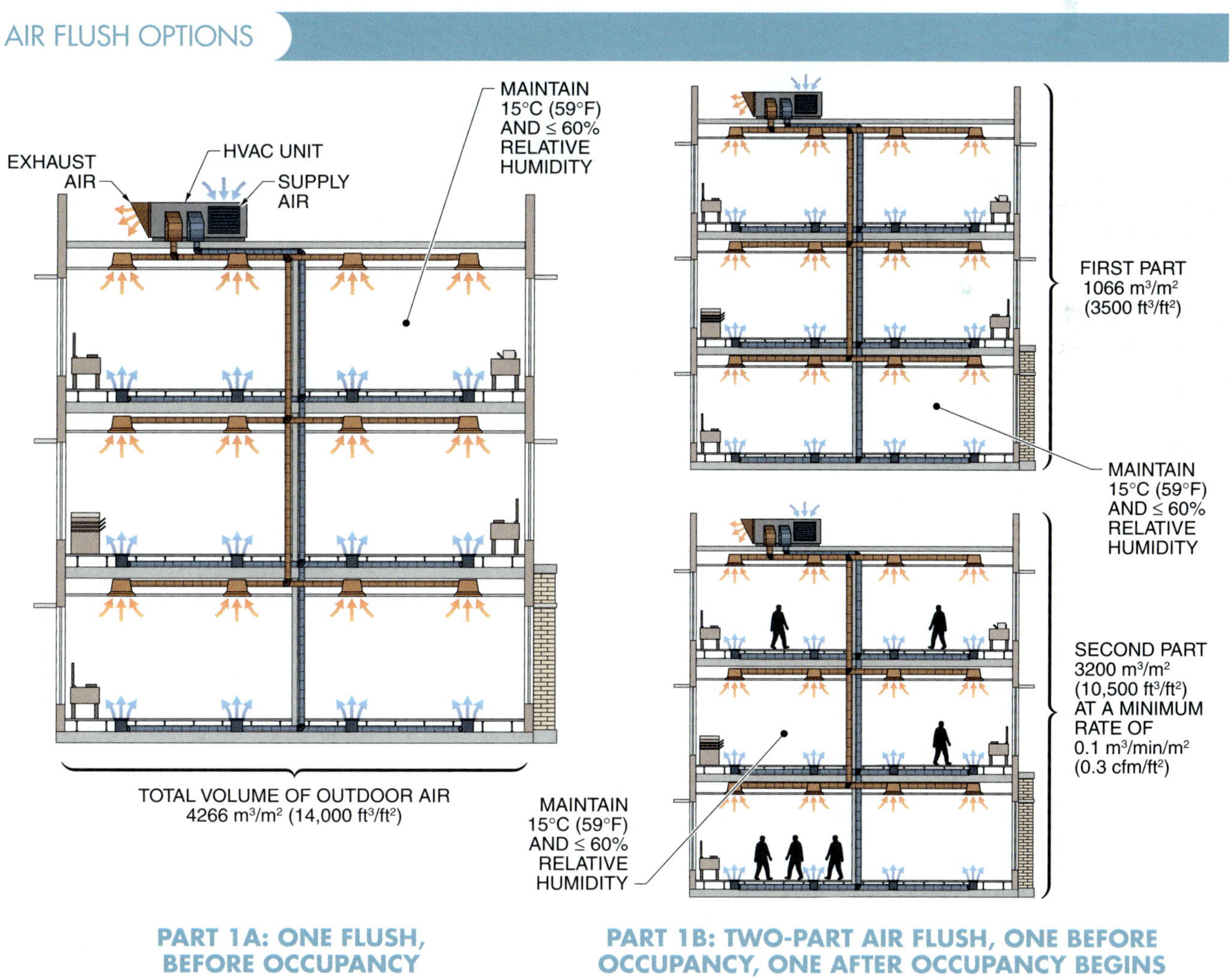

Figure 5-2. Feature 13, Air Flush, allows for an air flush to be performed entirely before occupancy or to be split up before and after occupancy, though each option has its own requirements.

FEATURE 16, HUMIDITY CONTROL

Proper humidity levels are important to the health and comfort of building occupants as well as the air quality in a building. When humidity is too low, building occupants may experience dryness and irritation of the eyes, throat, and mucous membranes. When humidity is too high, it allows for the accumulation and growth of microbial pathogens such as bacteria and mold. Excessive humidity can also lead to increased off-gassing of building materials.

Feature 16, Humidity Control, is an optimization for the New and Existing Interiors and New and Existing Buildings project types that requires either humidity control or humidity modeling. The local climate will have a significant impact on which option a project team decides to pursue.

BY THE NUMBERS

- *Humidity control can be important for indoor spaces because the off-gassing of formaldehyde can be 1.8 to 2.6 times higher when relative humidity increases by 35%.*

Part 1: Relative Humidity

Part 1: Relative Humidity of Feature 16 requires that one of two options be implemented for humidity control. Part 1a requires that the HVAC system have the ability to add or remove moisture to maintain the relative humidity within 30% and 50% at all times. Part 1b requires that the modeled humidity levels be maintained at 30% to 50% in the building for at least 95% of all business hours of the year.

FEATURE 17, DIRECT SOURCE VENTILATION

Typically, the best practice for preventing air pollutants from entering a space is to eliminate their source. Sources of air pollution may be building products (such as paints and cleaning supplies that contain VOCs), office equipment (such as printers and copiers that produce ozone), and specific rooms (such as bathrooms that may have high humidity leading to mold or mildew). However, since these building products, equipment, and spaces are necessary for proper building operations, they must be separated instead of eliminated. Feature 17, Direct Source Ventilation, is an optimization for all three project types that aims to separate harmful sources of air pollution from occupied spaces and to actively vent pollutants to the outside.

Part 1: Pollution Isolation and Exhaust

Part 1: Pollution Isolation and Exhaust of Feature 17 applies to cleaning and chemical storage units or spaces, all bathrooms, and all rooms that contain printers and copiers that do not meet the low-emission standards of Ecologo CCD 035, Blue Angel RAL-UZ 171, or Green Star. Ecologo CCD 05 and Blue Angel RAL-UZ 171 are voluntary international ecolabels for products that have undergone rigorous testing and auditing. Green Star is an international sustainability rating system for buildings. Part 1 requires that all spaces or storage units be closed from adjacent spaces with self-closing doors and the air inside the spaces or storage units be expelled from the spaces. Expelling or exhausting the air prevents the recirculation of pollutants.

FEATURE 18, AIR QUALITY MONITORING AND FEEDBACK

Feature 18, Air Quality Monitoring and Feedback, is an optimization for the New and Existing Interiors and New and Existing Buildings project types. This feature requires the monitoring of indoor air quality, the recording and reporting of data, and keeping the building managers and occupants informed of the building conditions in real time. The accurate monitoring and reporting of building conditions allows the adjustment of building systems to meet changing interior conditions and properly maintain positive air quality and occupant comfort.

Part 1: Indoor Air Monitoring

Part 1: Indoor Air Monitoring of Feature 18 requires that monitors take measurements of indoor pollutants and that the results be reported annually to the International WELL Building Institute (IWBI). At least one measurement per floor must be taken at least once per hour in a regularly occupied or common space. The monitors must collect and measure data at 1.2 m to 1.8 m (4′ to 6′) above the floor on two of the following pollutants:

- particle count (35,000 counts per m^3 [1000 counts per ft^3] or finer) or particle mass (10 $\mu g/m^3$ or finer)
- carbon dioxide (25 ppm or finer)
- ozone (10 ppb or finer)

Part 2: Air Data Record Keeping and Response

Part 2: Air Data Record Keeping and Response of Feature 18 requires that a written air monitoring policy be provided. Part 2a requires that the policy provide details of the monitoring and record keeping of the pollutants listed in Part 1: Indoor Air Monitoring. Part 2b requires that the records be kept for a minimum of three years. Part 2c requires that a plan be in place to address any unacceptable conditions.

Part 3: Environmental Measures Display

Part 3: Environmental Measures Display of Feature 18 requires that building occupants be able to see real-time environmental conditions. **See Figure 5-3.** For every 930 m^2 (10,000 ft^2) of occupied space, a real-time display that is at least 15 cm × 13 cm (5.9″ × 5.1″) must show temperature, humidity, and carbon dioxide concentration being monitored.

FEATURE 19, OPERABLE WINDOWS

The inclusion of operable exterior windows in a building's design allows building occupants some degree of control of their personal comfort. However, this must be weighed against the need to maintain high indoor air quality. Climatic conditions and outdoor air pollution levels affect whether natural ventilation strategies such as operable windows are able to remove indoor air pollutants. Prevailing breeze patterns can bring more pollutants and contaminants into a building than the operable windows let out. Feature 19, Operable Windows, is an optimization for all three project types that balances the occupants' comfort with the need to maintain indoor air quality.

ENVIRONMENTAL MEASURE DISPLAYS

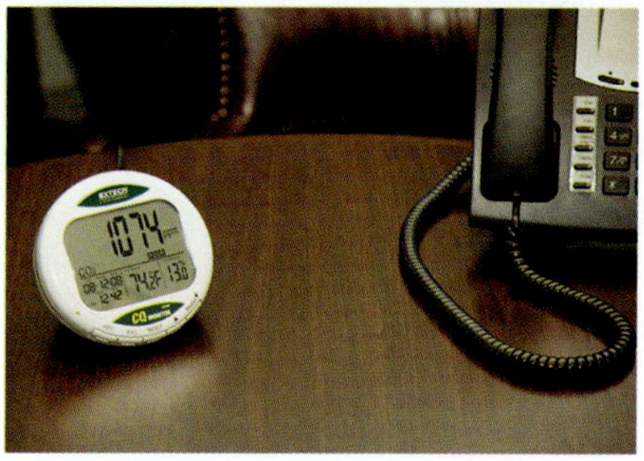

Extech Instruments

Figure 5-3. Part 3: Environmental Measures Display of Feature 18, Air Quality Monitoring and Feedback, requires the use of real-time environmental displays that monitor temperature, humidity, and carbon dioxide concentration.

Part 1: Full Control

Part 1: Full Control of Feature 19 simply requires operable windows in every regularly occupied space. These windows must allow building occupants to access outdoor air and daylight.

Part 2: Outdoor Air Measurement

Part 2: Outdoor Air Measurement of Feature 19 requires that an outdoor-air-measurement data collection system be located within 1.6 km (1 mile) of the building. This data collection system must collect readings for levels of ozone, particulate matter (PM_{10}), temperature, and humidity.

Part 3: Window Operation Management

Part 3: Window Operation Management of Feature 19 requires either software on the occupants' computers or smartphones or indicator lights on all operable windows that discourage occupants from opening the windows when the readings from Part 2 exceed one of the limits set in Part 3. These limits are as follows:

- ozone—51 ppb
- PM_{10}—50 μg/m^3
- temperature—±8°C (15°F) from set indoor building temperature
- humidity—60% relative humidity

FEATURE 20, OUTDOOR AIR SYSTEMS

Feature 20, Outdoor Air Systems, is an optimization for all three project types that allows the use of a dedicated outdoor air system. A *dedicated outdoor air system (DOAS)* is an HVAC system that provides 100% outside air directly to a building's zones for ventilation purposes. This type of HVAC system has better humidity control than other types of HVAC systems while supplying adequate outdoor air to all of the occupied spaces. DOASs also have increased air filtration and energy efficiency.

Part 1: Dedicated Outdoor Air Systems

Part 1: Dedicated Outdoor Air Systems of Feature 20 requires that DOASs meet one of two requirements. Part 1a requires that the system meet local codes or standards for DOASs. If there are no standards or the project cannot meet the local standards, Part 1b requires that a design review be performed by a qualified and registered professional mechanical engineer. The design review must address thermal comfort, ventilation rates, and system serviceability and reliability. The review must also ensure the DOAS's compliance with the ventilation standards used in Feature 03, Ventilation Effectiveness.

FEATURE 21, DISPLACEMENT VENTILATION

Feature 21, Displacement Ventilation, is an optimization for the New and Existing Interiors and New and Existing Buildings project types that addresses the design and performance of displacement ventilation. A *displacement ventilation system* is a ventilation system that introduces low-velocity supply air at a low level to displace warmer air that is then extracted at ceiling level. **See Figure 5-4.** This allows for the removal of pollutants that are concentrated toward the ceiling. The two types of displacement ventilation systems are low sidewall air distribution systems and underfloor air distribution (UFAD) systems.

Part 1: Displacement Ventilation Design and Application

Part 1: Displacement Ventilation Design and Application of Feature 21 requires that the displacement ventilation system for the project meet minimum design standards. For low sidewall air distribution systems, the System Performance Evaluation and ASHRAE Guidelines RP-949 should be used for the basis of design. For UFAD systems, the ASHRAE *UFAD Guide: Design, Construction and Operation of Underfloor Air Distribution Systems* should be used for the basis for design. In addition, Part 1 requires that the raised floor height allow for the underfloor area to be annually cleaned.

Part 2: System Performance

Part 2: System Performance of Feature 21 requires that a project using a displacement ventilation system meet two requirements. First, a computational fluid dynamics (CFD) analysis must be conducted for the system. The CFD analysis will provide a sophisticated evaluation on how air flows through the system and the associated energy transfer. The second requirement is that 75% of occupied space must meet the thermal comfort requirements of ASHRAE Standard 55-2013, *Thermal Environmental Conditions for Human Occupancy.*

DISPLACEMENT VENTILATION SYSTEMS

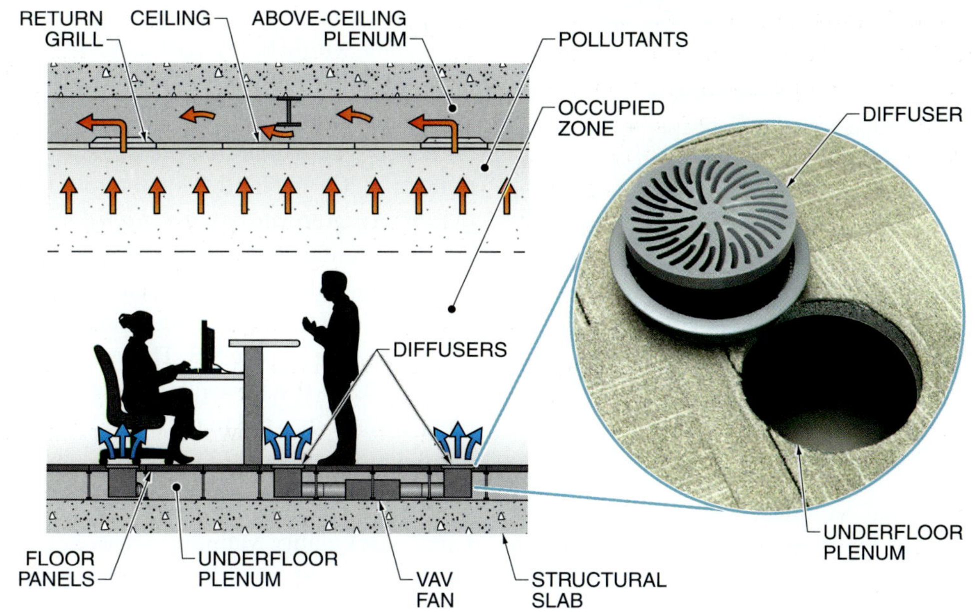

Figure 5-4. An underfloor air distribution (UFAD) system delivers fresh supply air near the occupant level, allowing air to flow from floor to ceiling and efficiently removing contaminants from the occupied zone.

FEATURE 22, PEST CONTROL

Feature 22, Pest Control, is an optimization for the New and Existing Interiors and New and Existing Buildings project types. Not only can pests be carriers of disease, their bodies, feces, and salvia contain allergens that can trigger asthma and allergic reactions. This feature requires strategies to be implemented for reducing the unhygienic conditions that appeal to pests and inspections for ensuring there are no signs of infestation.

Part 1: Pest Reduction

Part 1: Pest Reduction of Feature 22 addresses strategies for reducing the chances of pest infestation. These strategies require the following:

- Perishable foods not in a refrigerator must be stored in sealed containers.
- Indoor garbage cans less than 113 L (30 gal.) must have lids that can be operated without the use of hands or be enclosed by cabinetry in an undercounter, pull-out drawer that has a handle separate from the garbage can.
- Indoor garbage cans greater than 113 L (30 gal.) must have a lid.

BY THE NUMBERS

- *Up to 60% of asthmatic people who live in cities or other urban environments also have reactions to cockroach allergens.*

Part 2: Pest Inspection

Part 2: Pest Inspection of Feature 22 requires that visual inspections be performed. The inspections must show that there are no signs of cockroach, termite, or other pest infestations.

FEATURE 23, ADVANCED AIR PURIFICATION

Feature 23, Advanced Air Purification, is an optimization for all three project types that expands on the requirements of Feature 05, Air Filtration—an Air Concept precondition. In certain locations, the quality of outdoor air may be significantly lower than desired for providing optimal indoor air quality. In order to keep indoor air quality at acceptable levels, this feature addresses advanced methods of air filtration, air sanitation, and the maintenance of air filtration and sanitization systems.

Part 1: Carbon Filtration

Part 1: Carbon Filtration of Feature 23 requires that buildings with recirculated air use one of two carbon filtration methods to reduce the amounts of VOCs in the air. One method is to install activated carbon filters or combination particulate/carbon filters in ductwork as a part of the main HVAC system. The other method is to install properly sized, standalone air purifiers with carbon filters in all regularly occupied spaces.

Part 2: Air Sanitization

Part 2: Air Sanitization of Feature 23 applies to spaces with more than 10 regular occupants, within buildings that recirculate air. Part 2 requires that either an ultraviolet germicidal irradiation (UVGI) system or a photocatalytic oxidation system be used to remove VOCs and harmful pathogens from the air. A *photocatalytic oxidation system* is an air sanitation system that uses a UV light along with a catalyst, usually titanium dioxide, to break down contaminants in the air stream. **See Figure 5-5.** These air sanitization systems can be standalone systems or be integrated into the main HVAC system.

Part 3: Air Quality Maintenance

Part 3: Air Quality Maintenance of Feature 23 requires that any filtration or sanitization system be properly maintained per the manufacturer's recommendations. A record of this maintenance must be submitted to IWBI annually. The maintenance records must show that the chosen system continues to properly operate.

PHOTOCATALYTIC OXIDATION SYSTEMS

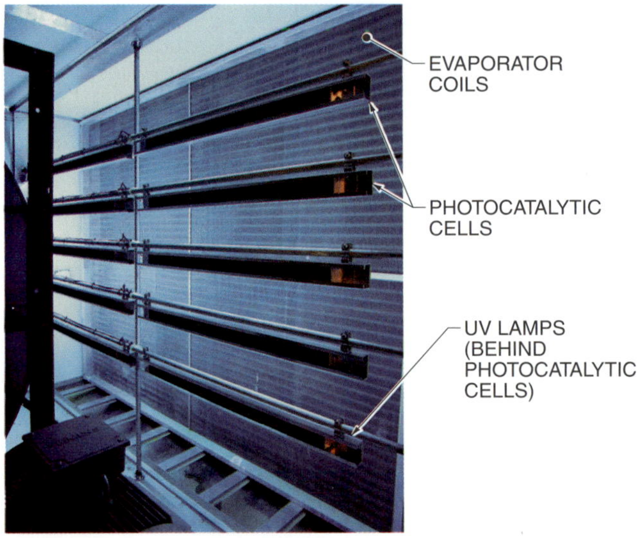

Figure 5-5. A photocatalytic oxidation system uses a UV light along with a catalyst, usually titanium dioxide, to break down contaminants in the air stream.

FEATURE 24, COMBUSTION MINIMIZATION

The incomplete combustion of fuels can lead to the production of carbon monoxide (CO) and particulates in the form of soot. CO is dangerous because it replaces oxygen in the hemoglobin of red blood cells, which limits the ability of blood to deliver oxygen and can lead to hypoxia and possibly death. Symptoms of hypoxia include headaches, nausea, and loss of consciousness.

The first part of Feature 24, Combustion Minimization, is an optimization for all three project types, but the second, third, and fourth parts only apply to the Core and Shell and New and Existing Building types. The aim of this feature is to reduce building occupant exposure to combustion-related air pollution by prohibiting, regulating, and controlling the sources of interior and exterior combustion.

Part 1: Appliance and Heater Combustion Ban

Part 1: Appliance and Heater Combustion Ban of Feature 24 forbids combustion-based equipment in regularly occupied spaces. Examples of combustion-based equipment include fireplaces, stoves, space-heaters, ranges, and ovens.

BY THE NUMBERS

- *The affinity for carbon monoxide to bind to the hemoglobin of red blood cells is 210 times stronger than that of oxygen, meaning it is 210 times more likely to be carried through the bloodstream than oxygen.*
- *Carbon monoxide from nonvehicle sources such as fuel-burning appliances and engine-powered equipment is responsible for an estimated 170 deaths per year in the United States.*

Part 2: Low-Emission Combustion Sources

Part 2: Low-Emission Combustion Sources of Feature 24 regulates combustion-based equipment that is used for heating, cooling, water heating, process heating, or power generation in a building. The equipment must meet California's South Coast Air Quality Management District (SCAQMD) rules for pollution. **See Figure 5-6.** Specific pieces of combustion-based equipment that are regulated by Part 2 include internal combustion engines; furnaces; boilers, steam generators, and process heaters; and water heaters.

LOW-EMISSION COMBUSTION SOURCES

Figure 5-6. Combustion-based equipment, such as a boiler, that is installed in a building must meet California's SCAQMD rules for pollution and have low emissions.

Part 3: Engine Exhaust Reduction

Part 3: Engine Exhaust Reduction of Feature 24 prohibits the idling of vehicles for more than 30 seconds. Signage that states this 30 second rule must be visible to building occupants and visitors in pick-up, drop-off, and parking areas.

Part 4: Construction Equipment

Part 4: Construction Equipment of Feature 24 requires that efforts be made to reduce the amount of particulate matter (PM) produced by diesel vehicles and construction equipment on a project site for construction occurring within one year prior to perfromance verification. Part 4 applies to both nonroad and on-road diesel vehicles and equipment. The following requirements must be met:

- All nonroad diesel vehicles or equipment must comply with U.S. EPA Tier 4 PM emissions standards or a local equivalent. When the equipment is first delivered to the project site, it may be retrofitted with emission-reducing technology that has EPA or California Air Resources Board (CARB) approval.
- All on-road vehicles and equipment must meet the requirements of U.S. EPA 2007 standards for PM or a local equivalent. This equipment may also be retrofitted with emission-reducing technology that has EPA or CARB approval when it is first delivered to the project site.
- Both nonroad and on-road vehicles and equipment should be operated, loaded, and unloaded away from the air intakes and openings of buildings that are adjacent to the project.

FEATURE 25, TOXIC MATERIAL REDUCTION

Toxic materials that can adversely affect human health include chemical compounds such as perfluorinated compounds (PFCs), polybrominated diphenyl ethers (PBDEs), phthalates, isocyanate-based polyurethane, and urea-formaldehyde (UF). Feature 25, Toxic Material Reduction, is an optimization for the New and Existing Interiors and New and Existing Buildings project types that limits the amount of hazardous compounds allowed as ingredients in the building materials and products installed inside the weatherproofing membrane of the project.

Part 1: Perfluorinated Compound Limitation

Perfluorinated compounds (PFCs) are a family of fluorine-containing chemicals with unique properties to make materials stain- and stick-resistant. These chemicals can be found in drapes, interior furniture and furnishings, and carpets. PFCs can affect the immune system, cause developmental and reproductive issues, increase cholesterol levels, and increase the risk of cancer. Part 1: Perfluorinated Compound Limitation of Feature 25 prohibits the use of furniture or furnishings if the levels of PFCs are equal to or greater than 100 ppm in the components that constitute at least 5% by weight of the furniture or furnishing assemblies.

Part 2: Flame Retardant Limitation

A *halogenated flame retardant* is a chemical bonded with one of the halogen elements, such as chlorine or bromine, used in thermoplastics, thermosets, textiles, and coatings that inhibits or resists the spread of fire. Halogenated flame retardants, such as polybrominated diphenyl ethers, are used in building products.

Polybrominated diphenyl ethers (PBDEs) are a group of brominated hydrocarbons that are used as flame retardants for plastics, foams, furniture and furnishings, textiles, and other household products. PBDEs can disrupt the endocrine system, impair the immune system, cause developmental and behavioral problems, and increase the risk of cancer. Part 2: Flame Retardant Limitation of Feature 25 limits halogenated flame retardants such as PBDEs to 0.01% (100 ppm) in the following products:

- window and waterproofing membranes, door and window frames, and siding
- flooring, ceiling tiles, and wall coverings

- piping and electrical cables, conduits, and junction boxes
- sound and thermal insulation
- upholstered furniture and furnishings, textiles, and fabrics

Part 3: Phthalate (Plasticizers) Limitation

Phthalates are a group of chemicals used to make plastics more flexible and harder to break. Several types of phthalates have been banned from children's toys and other child-specific products, but these chemicals can still be found in a variety of building and consumer products. Part 3: Phthalate (Plasticizers) Limitation of Feature 25 limits the amount of phthalates to 0.01% (100 ppm) in the following products:

- flooring and carpet
- wall coverings, window blinds and shades, shower curtains, furniture, and upholstery
- plumbing pipes and moisture barriers

Part 4: Isocyanate-Based Polyurethane Limitation

Isocyanate-based polyurethane is used as an interior finish. *Polyurethane* is a synthetic resin used chiefly in paints and varnishes. *Isocyanate* is an organic compound that is used in surface finishes and coatings. Exposure to isocyanates through polyurethane products can cause irritation of the skin and mucous membranes, chest tightness, and difficulty breathing. Part 4: Isocyanate-Based Polyurethane Limitation of Feature 25 prohibits the use of isocyanate-based polyurethane in interior finishes.

Part 5: Urea-Formaldehyde Restriction

Urea-formaldehyde is a low-cost thermosetting resin that is used in the wood product industry. Urea-formaldehyde is also used in certain insulations and adhesives. Exposure to urea-formaldehyde can cause headaches and mucous membrane irritation. Part 5: Urea-Formaldehyde Restriction of Feature 25 limits the levels of urea-formaldehyde to 100 ppm for the following products:

- furniture or any composite wood products
- laminating adhesives and resins
- thermal insulation

FEATURE 26, ENHANCED MATERIAL SAFETY

Feature 26, Enhanced Material Safety, is an optimization for the New and Existing Interiors and New and Existing Buildings project types. This feature encourages the use of materials that are free of potentially harmful ingredients and are, therefore, healthier for building occupants and manufacturing and maintenance workers. Project teams should choose materials that meet the requirements of one or more voluntary material ingredient reporting programs.

Part 1: Precautionary Material Selection

Part 1: Precautionary Material Selection of Feature 26 requires that at least 25% (by cost) of all furnishings, built-in furniture, interior finishes, and finish materials for the project meet one or more requirements from the listed voluntary material reporting programs. These voluntary material reporting programs include the following:

- One of three labels from the International Living Future Institute, either a Declare: Living Building Challenge Red List Free label; a Declare: Living Building Challenge Compliant label; or a Living Product Challenge label
- Cradle to Cradle™ Material Health Certified with a V2 Gold or Platinum or V3 Bronze, Silver, Gold, or Platinum Material Health Score
- No GreenScreen® Benchmark 1, List Translator 1, or List Translator Possible 1 substances over 1000 ppm, as verified by a qualified professional

FEATURE 27, ANTIMICROBIAL ACTIVITY FOR SURFACES

Feature 27, Antimicrobial Activity for Surfaces, is an optimization for the New and Existing Interiors and New and Existing Buildings project types that requires high-touch surfaces to be antimicrobial, abrasion-resistant, and nonleaching. Certain materials, such as antimicrobial coatings and copper, are capable of reducing the amount of pathogens on surfaces without leaching significant amounts of antibacterial substances into the surrounding environment. **See Figure 5-7.** Another method to kill pathogens and microorganisms on surfaces is the use of short-wavelength ultraviolet (UV-C) light. UV-C light disrupts the DNA structure of pathogens and microorganisms, killing them and preventing them from reproducing.

WELLNESS FACT

Both pure copper and copper alloys, such as brasses, bronzes, and copper-nickels, are antimicrobial. Pure copper or copper alloy surfaces are best because copper plating and coatings may wear or scratch too easily, providing a place for pathogens to survive.

Part 1: High-Touch Surfaces

Part 1: High-Touch Surfaces of Feature 27 requires that one of two antimicrobial strategies be used on high-touch surfaces. The high-touch surfaces in a building include all countertops and fixtures in bathrooms and kitchens, handles, doorknobs, light switches, and elevator buttons. These surfaces must be coated with or consist of abrasion-resistant, nonleaching materials that meet EPA requirements for antimicrobial activity, or they must be cleaned per the manufacturer's instructions with a UV cleaning device.

FEATURE 28, CLEANABLE ENVIRONMENT

Feature 28, Cleanable Environment, is an optimization for the New and Existing Interiors and New and Existing Buildings project types that addresses the material properties of high-touch surfaces and the cleanability of occupied spaces. Specifying easy-to-clean materials for high-touch surfaces can help prevent the spread of pathogens and other dangerous substances. This feature expands on the requirements of

ANTIMICROBIAL SURFACES

CuVerro® (Olin Brass)

Figure 5-7. Certain materials, such as antimicrobial coatings and copper, are capable of reducing the amount of pathogens on surfaces without leaching significant amounts of antibacterial substances into the surrounding environment.

Feature 09, Cleaning Protocol, which is a precondition for New and Existing Interiors and New and Existing Buildings projects.

Part 1: Material Properties

Part 1: Material Properties of Feature 28 lists the requirements for high-touch, nonporous surfaces found in a building. A project team should have already created a list of high-touch surfaces within their project as required by Feature 09, Cleaning Protocol. A list of common high-touch surfaces can also be found in Table A1 of Appendix C of the WELL Building Standard. Part 1 lists the following three requirements that high-touch, nonporous surfaces in a project must meet:

- smooth and free of visible defects that can make the surface difficult to clean
- smooth welds and joints to allow for easy cleaning
- no crevices or other hard-to-reach places that can trap dirt or pathogens

High-touch surfaces such as faucets and sinks must be smooth, corrosion-resistant, and easily cleaned.

Part 2: Cleanability

Part 2: Cleanability of Feature 28 encourages the design of spaces to allow for easy cleaning of high-touch surfaces. Part 2 requires the following design strategies:

- If flooring products such as rugs and carpet tiles are necessary, they must be easy to clean or removable and must specifically exclude wall-to-wall carpet.
- Adequate storage space must be provided for movable items that normally occupy a space so that the high-touch surfaces can be thoroughly accessed and cleaned.
- The joints (90° angles) where the walls meet windows and floors must be sealed to eliminate difficult-to-clean gaps that can collect pathogens or toxins.

FEATURE 29, CLEANING EQUIPMENT

Once a building is constructed, the spaces within the building must be properly cleaned to maintain a healthy environment. The use of high-performance cleaning equipment can aid in this process. Feature 29, Cleaning Equipment, is an optimization for the New and Existing Interiors and New and Existing Buildings project types that addresses the cleaning equipment and products used to maintain building spaces as well as proper storage for any toxic cleaning chemicals.

Part 1: Equipment and Cleaning Agents

Part 1: Equipment and Cleaning Agents of Feature 29 requires that cleaning equipment and products meet minimum requirements. These requirements are as follows:

- Any mops, rags, and dusters used for cleaning nonporous surfaces must be made of microfiber materials. The size of the microfiber cannot be any larger than 1.0 denier. Denier is the weight in grams of a 9000 m length of fiber; the smaller the number, the thinner the fiber. **See Figure 5-8.**
- Mops must be a type that does not need to be wrung by hand.
- Vacuum cleaners must contain high-efficiency particulate air filters. A *high-efficiency particulate air (HEPA)* filter is a filter that removes 99.97% of all particles greater than 0.3 mm and satisfies standards of efficiency set by the Institute of Environmental Sciences and Technology (IEST).

MICROFIBER CLEANING PRODUCTS

Figure 5-8. Part 1: Equipment and Cleaning Agents of Feature 29, Cleaning Equipment, requires that any mops, rags, and dusters used for cleaning nonporous surfaces be made of microfiber materials no larger than 1.0 denier.

WELLNESS FACT

Ammonia is commonly used for degreasing surfaces. Ammonia-based cleaners can burn or irritate living tissue. Bleach is often used as an antibacterial and antifungal disinfectant to sanitize surfaces. Overexposure to bleach can cause skin and respiratory irritation. If mixed, bleach and ammonia can cause severe lung damage or death.

Part 2: Chemical Storage

Part 2: Chemical Storage of Feature 29 addresses the storage of cleaning chemicals, specifically those containing bleach and ammonia. Dangerous gases will be released if a bleach-containing product is accidently mixed with an ammonia-based product. Bleach can also react with other acidic cleaners. Part 2 requires that bleach and ammonia-based products be kept in separate bins in a cleaning storage area. Part 2 also requires that any bins or bottles containing these products have warning labels that specify that they must not be mixed.

KEY TERMS AND DEFINITIONS

air exfiltration: The movement of air out of a conditioned building space.

air flushing: A technique used to remove or reduce airborne contaminants and pollutants by running the HVAC system for an extended period of time after construction is complete but before occupancy.

air infiltration: The movement of air into a conditioned building space through an unwanted void in the building envelope.

dedicated outdoor air system (DOAS): An HVAC system that provides 100% outside air directly to a building's zones for ventilation purposes.

displacement ventilation system: A ventilation system that introduces low-velocity supply air at a low level to displace warmer air that is then extracted at ceiling level.

halogenated flame retardant: A chemical bonded with one of the halogen elements, such as chlorine or bromine, used in thermoplastics, thermosets, textiles, and coatings that inhibits or resists the spread of fire.

high-efficiency particulate air (HEPA) filter: A filter that removes 99.97% of all particles greater than 0.3 μm and satisfies standards of efficiency set by the Institute of Environmental Sciences and Technology (IEST).

isocyanate: An organic compound that is used in surface finishes and coatings.

perfluorinated compounds (PFCs): A family of fluorine-containing chemicals with unique properties that make materials stain- and stick-resistant.

photocatalytic oxidation system: An air sanitation system that uses a UV light along with a catalyst, usually titanium dioxide, to break down contaminants in the air stream.

KEY TERMS AND DEFINITIONS *(continued)*

phthalates: A group of chemicals used to make plastics more flexible and harder to break.

polybrominated diphenyl ethers (PBDEs): A group of brominated hydrocarbons that are used as flame retardants for plastics, foams, furniture and furnishings, textiles, and other household products.

polyurethane: A synthetic resin used chiefly in paints and varnishes.

urea-formaldehyde: A low-cost thermosetting resin that is used in the wood product industry.

CHAPTER REVIEW

Completion

______________ **1.** The majority of the optimizations for the Air Concept are found in Feature 13 through Feature ___.

______________ **2.** The interior temperature and humidity of a building must be maintained at a minimum of 15°C (59°F) and no more than ___% relative humidity for an air flush.

______________ **3.** Feature 14, ___, is an optimization for all three project types that requires building envelope commissioning and a plan for action and remediation.

______________ **4.** In order to meet the requirements of Part 1: Increased Outdoor Air Supply of Feature 15, Increased Ventilation, the ventilation supply rates must exceed the rates met in Feature 03, Ventilation Effectiveness, by ___%.

______________ **5.** Part 1: ___ of Feature 17, Direct Source Ventilation, requires that all spaces or storage units be closed from adjacent spaces with self-closing doors and the air inside the spaces or storage units be expelled from the spaces.

______________ **6.** Part 2: Air Data Record Keeping and Response of Feature 18, Air Quality Monitoring and Feedback, requires that records be kept for a minimum of ___ years.

______________ **7.** Feature 19, Operable Windows, is an optimization for ___ project type(s).

______________ **8.** A dedicated outdoor air system (DOAS) is an HVAC system that provides ___% outside air to a building's zones for ventilation purposes.

______________ **9.** The second requirement for Part 2: System Performance of Feature 21, Displacement Ventilation, requires that 75% of occupied space meet the thermal comfort requirements of ASHRAE Standard ___, *Thermal Environmental Conditions for Human Occupancy.*

______________ **10.** Part 2: Air Sanitization of Feature 23, Advanced Air Purification, applies to spaces with more than ___ regular occupants, within buildings that recirculate air.

______________ **11.** Part 3: Engine Exhaust Reduction of Feature 24, Combustion Minimization, requires signage that prohibits the idling of vehicles for more than ___ seconds.

______________ **12.** Part 4: Isocyanate-Based Polyurethane Limitation of Feature 25, Toxic Material Reduction, prohibits the use of isocyanate-based polyurethane in ___.

CHAPTER REVIEW *(continued)*

____________________ **13.** To meet Part 1: Precautionary Material Selection of Feature 26, Enhanced Material Safety, at least ___% (by cost) of all furnishings, built-in furniture, interior finishes, and finish materials for the project must meet one or more requirements from the listed voluntary material reporting programs.

____________________ **14.** Per Part 1: High-Touch Surfaces of Feature 27, ___, high-touch surfaces must be cleaned per the manufacturer's instructions with a UV cleaning device.

____________________ **15.** Any mops, rags, and dusters used for cleaning nonporous surfaces must be made of microfiber materials that cannot be any larger than ___ denier.

____________________ **16.** Part 2: Chemical Storage of Feature 29, Cleaning Equipment, addresses the storage of cleaning chemicals, specifically those containing bleach and ___.

Short Answer

1. What are the effects of humidity that is too low and humidity that is too high?

2. List the three pollutants from which two may be selected for measuring and monitoring per Part 1: Indoor Air Monitoring of Feature 18, Air Quality Monitoring and Feedback.

3. List the four parameters that must be monitored by a data collection system per Part 2: Outdoor Air Measurement of Feature 19, Operable Windows, as well as their limits for open windows per Part 3: Window Operation Management.

4. What qualified and registered professional must review the design of a dedicated outdoor air system (DOAS) and what must the review address?

5. List the strategies for reducing the chances of pest infestations.

CHAPTER REVIEW *(continued)*

6. What are the two methods of using carbon filtration to reduce VOCs in indoor air?

7. List the five pieces of combustion-based equipment that are forbidden in regularly occupied spaces.

8. List the types of combustion-based equipment used for heating, cooling, water heating, process heating, or power generation in a building that must meet California's South Coast Air Quality Management District (SCAQMD) rules for pollution.

9. Which four parts of Feature 25, Toxic Material Reduction, limit the amounts of different hazardous compounds to 100 ppm in the building materials and products installed inside the weatherproofing membrane of the project?

10. List the requirements from the voluntary material reporting programs that can be used to meet Part 1: Precautionary Material Selection of Feature 26, Enhanced Material Safety.

11. Which high-touch surfaces in a building must be coated with or consist of abrasion-resistant, nonleaching materials that meet EPA requirements for antimicrobial activity per Part 1: High-Touch Surfaces of Feature 27, Antimicrobial Activity for Surfaces?

12. List the three requirements that the high-touch surfaces in a project must meet to achieve Part 1: Material Properties of Feature 28, Cleanable Equipment.

WELL AP EXAM PRACTICE QUESTIONS

1. A finished project that is not yet occupied maintains an indoor temperature of 17°C (62.5°F) and 55% relative humidity. What is the total volume (in m^3) of outdoor air per m^2 of floor area that must be used for an air flush if the project team wishes to complete the entire air flush before occupancy for Feature 13, Air Flush?
   A. 1066
   B. 3200
   C. 4266
   D. 5600

2. Between which two percentages must an HVAC unit maintain relative humidity or modeled humidity for Part 1: Relative Humidity of Feature 16, Humidity Control?
   A. 30% to 50%
   B. 35% to 55%
   C. 40% to 60%
   D. 50% to 80%

3. Which optimization feature in the Air Concept includes a requirement that 75% of all regularly occupied space meet the thermal comfort requirements of ASHRAE Standard 55-2013, *Thermal Environmental Conditions for Human Occupancy*?
   A. Feature 15, Increased Ventilation
   B. Feature 17, Direct Source Ventilation
   C. Feature 19, Operable Windows
   D. Feature 21, Displacement Ventilation

4. It has been ensured that chemical compounds such as PFCs, PBDEs, phthalates, isocyanate-based polyurethanes, and urea-formaldehyde are absent or at levels under 100 ppm in the installed building materials and products of a New and Existing Interiors project. Which optimization feature in the Air Concept can this project achieve?
   A. Feature 07, Construction Pollution Management
   B. Feature 11, Fundamental Material Safety
   C. Feature 25, Toxic Material Safety
   D. Feature 26, Enhanced Material Safety

5. Which methods of documentation and performance verification must a project use to prove that the requirements of Part 2: Cleanability of Feature 28, Cleanable Environment, have been fulfilled?
   A. A letter of assurance from the owner and an operations schedule as an annotated document
   B. A professional narrative as an annotated document and a visual inspection as an on-site assessment
   C. A letter of assurance from an MEP engineer and a professional narrative as an annotated document
   D. Letters of assurance from the architect and contractor and a spot check as an on-site assessment

CHAPTER SIX

WATER

Clean drinking water is essential for human life since water is the principal component of the human body and is responsible for many bodily functions. However, water is a precious commodity since it is scarce in many areas of the world. Even if water is available, it may be polluted, contaminated, or contain harmful pathogens. The primary goal of the Water Concept in the WELL Building Standard is to ensure that clean water is provided for building occupants.

OBJECTIVES

- Identify the features in the Water Concept as preconditions or optimizations.
- Describe the minimum standards for fundamental water quality.
- Identify the inorganic contaminants that can affect water quality and their limits.
- Identify the organic contaminants that can affect water quality and their limits.
- Identify the agricultural contaminants that can affect water quality and their limits.
- Identify the public water additives that can affect water quality and their limits.
- Describe the different processes for water filtration and the minimum requirements for water testing.
- Describe recommended practices for promoting drinking water access.

KEY TERMS

- agricultural contaminant
- atrazine
- chloramine
- chlorine
- coliform
- disinfectant
- disinfectant by-product (DBP)
- fertilizer
- fluoride
- glyphosate
- granular activated carbon (GAC) filtration system
- haloacetic acid (HAA)
- herbicide
- inorganic contaminant
- kinetic degradation fluxion (KDF) filter
- Legionella
- microbial cyst
- nephelometric turbidity unit (NTU)
- nonpotable water
- organic contaminant
- pathogen
- pesticide
- potable water
- process water
- reverse-osmosis (RO) filtration system
- simazine
- trihalomethane (THM)
- turbidity
- 2,4-dichlorophenoxyacetic acid (2,4-D)
- ultraviolet germicidal irradiation (UVGI)

Learner Resources

ATPeResources.com/QuickLinks
Access Code: 935534

WATER CONCEPT

Water is so important to the human body that a person can live weeks without food but only days without water. The quality of water that is consumed by the human body is also important. Water has the ability to carry a variety of unseen dangers. Water contaminated by heavy metals, pathogens, or other chemicals can have harmful short-term and long-term effects on the human body.

There are four routes of exposure that allow water contaminated by heavy metals, chemical substances, or pathogens to enter the human body. The four routes of exposure are ingestion, inhalation, skin absorption, and injection. The Water Concept mainly addresses the quality of water that is being ingested, though it also applies to the quality of water that may be inhaled or absorbed through the skin.

The Water Concept in WELL includes eight features that promote water testing, treatment, and filtration to ensure that any water ingested or contacted by building occupants is clean and free of contaminants. Core and Shell projects must meet five preconditions, but two optimizations can be pursued for higher certification. New and Existing Interiors and New and Existing Buildings projects must meet five preconditions as well, but three optimizations can be pursued for higher certification. **See Figure 6-1.**

WATER SUPPLY DISTRIBUTION

Water is primarily supplied to a building from sources such as surface water (rivers, lakes, and reservoirs) or municipal wells. First, water supply distribution utilities draw water from these sources. The water is filtered and treated to meet federal regulations and standards or other applicable laws. Then, the filtered and treated water is stored for later use or pumped to a building through a system of water supply pipes.

In rural or less-accessible locations, buildings may be supplied with water through private wells. Private wells are not regulated by federal law, but there may be state or local regulations concerning the quality of well water. Any treatment of private well water occurs where the water enters a building or at the point of use.

Water Types

Three types of water can be used in, or on the site of, a building: potable, nonpotable, and process water. *Potable water* is water that is fit for human consumption. Potable water can be used for any function within a building, but it is the only type of water that has been made safe for drinking. *Nonpotable water* is water that is not fit for human consumption. *Process water* is water that is used for cooling towers, boilers, and industrial processes. Process water can be either potable or nonpotable water. Whichever water type is used, the process water often undergoes filtration and treatment such as softening and demineralization. If potable water is used for process water, it is no longer safe for drinking.

The Water Concept mainly addresses the quality of potable water. Although potable water is safe for drinking, it can still negatively affect human health. The Water Concept narrows the scope of potable water by differentiating between water that is designated for human contact and water that is designated for human consumption.

Potable water that is designated for human contact not only includes the water in water closets and urinals but also any water that may be consumed. However, potable water that is specifically designated for human consumption is held to stricter standards than the water designated for human contact. Potable water for human consumption includes the water used for drinking, bathing, cooking, dishwashing, oral hygiene, ice-making, and food production.

The water supply to a building can be contaminated at the source, in the supply lines, at the building connection, in the building piping, or inside the plumbing fixtures at the point of use. Treatment plants may not always be able to completely remove all of the health dangers from water that has been taken from surface sources and wells. Also, incoming water can leach certain compounds, such as lead, from the supply piping as the water moves from the treatment plant to the building. The water may also pick up contaminants from plumbing fixtures, or it may promote the growth of pathogens if it remains stagnant too long.

WELL Building Standard Features: Water Concept

Feature		Project Type			Verification Documentation		
		Core and Shell	New and Existing Interiors	New and Existing Buildings	Letter of Assurance	Annotated Documents	On-Site Assessment
30, Fundamental Water Quality							
	Part 1: Sediment	P	P	P			Performance Test
	Part 2: Microorganisms	P	P	P			Performance Test
31, Inorganic Contaminants							
	Part 1: Dissolved Metals	P	P	P			Performance Test
32, Organic Contaminants							
	Part 1: Organic Pollutants	P	P	P			Performance Test
33, Agricultural Contaminants							
	Part 1: Herbicides and Pesticides	P	P	P			Performance Test
	Part 2: Fertilizers	P	P	P			Performance Test
34, Public Water Additives							
	Part 1: Disinfectants	P	P	P			Performance Test
	Part 2: Disinfectant By-products	P	P	P			Performance Test
	Part 3: Fluoride	P	P	P			Performance Test
35, Periodic Water Quality Testing							
	Part 1: Quarterly Testing	–	O	O		Operations Schedule	
	Part 2: Water Data Record Keeping and Response	–	O	O		Operations Schedule	
36, Water Treatment							
	Part 1: Organic Chemical Removal	O	O	O	MEP		Spot Check
	Part 2: Sediment Filter	O	O	O	MEP		Spot Check
	Part 3: Microbial Elimination	O	O	O	MEP		Spot Check
	Part 4: Water Quality Maintenance	O	O	O		Operations Schedule	
	Part 5: Legionella Control	O	O	O		Professional Narrative	
37, Drinking Water Promotion							
	Part 1: Drinking Water Taste Properties	O	O	O			Performance Test
	Part 2: Drinking Water Access	–	O	O	Architect		Spot Check
	Part 3: Water Dispenser Maintenance	–	O	O		Operations Schedule	

Figure 6-1. The Water Concept in WELL includes eight features that promote water testing, treatment, and filtration to ensure that any water ingested or contacted by building occupants is clean and free of contaminants.

FEATURE 30, FUNDAMENTAL WATER QUALITY

Feature 30, Fundamental Water Quality, is the starting point for all other preconditions and optimizations in the Water Concept. This feature is a precondition for all three project types. All water that is delivered to the project area, except water that will not come into contact with humans, must meet minimum standards for clarity and coliform levels.

BY THE NUMBERS

- *The World Health Organization (WHO) reports that almost one billion people lack access to safe drinking water worldwide and that two million annual deaths can be attributed to unsafe water, sanitation, and hygiene.*

Part 1: Sediment

Water clarity is measured in turbidity. *Turbidity* is the amount of cloudiness in a liquid caused by suspended solids that are usually invisible to the naked eye. The higher the turbidity, the greater the amount of sedimentation or other substances (such as pathogens feeding off of the turbid substances) contained within the water sample.

Part 1: Sediment of Feature 30 requires that a water sample from a project site have a turbidity level of less than 1.0 nephelometric turbidity units (NTU). A *nephelometric turbidity unit (NTU)* is the unit of measure for the turbidity of water. The maximum turbidity for public drinking water should not exceed 5.0 NTU. However, most water utilities attempt to maintain a turbidity of 0.1 NTU.

Low turbidity is important because it allows certain water purification methods, such as ultraviolet (UV) light treatment, to work efficiently. When the turbidity of water is high, UV rays are blocked by the sedimentation and cannot reach any microorganisms that may be present. Turbidity can be tested with laboratory equipment or with a portable turbidity meter. **See Figure 6-2.**

TURBIDITY METERS

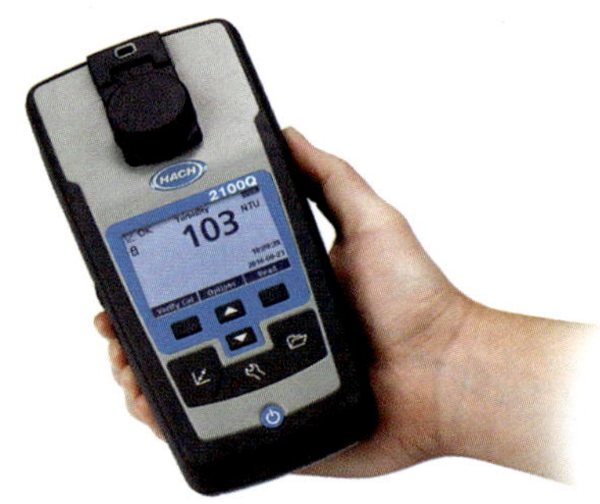

Hach Company

Figure 6-2. Turbidity, the amount of cloudiness in a liquid caused by suspended solids that are invisible to the naked eye, can be measured on site with a portable turbidity meter or in a laboratory.

Part 2: Microorganisms

Along with the turbidity test and measurements, a water sample must also be tested to ensure that there is no presence of coliforms. A *coliform* is a microorganism that includes bacteria such as E. coli. Coliforms are naturally found in soil, vegetation, and in the intestinal tract of mammals, including humans. Although a few strains of coliform bacteria may cause serious illness, most do not cause harm.

Increased levels of coliforms may be an indicator of other pathogens in the water. A *pathogen* is an infectious biological agent such as a bacterium, virus, or fungus that is capable of causing disease in its host. Part 2: Microorganisms of Feature 30 requires that a water sample be tested for coliforms. Total coliforms, which includes coliforms from all sources, must not be detected in the water sample. The total coliform test can adequately test the effectiveness of a project's water filtration.

FEATURE 31, INORGANIC CONTAMINANTS

An *inorganic contaminant* is an element or compound that may be found in a water supply, occurring from natural sources such as the geology of a location, resulting from human activities such as mining and industry, or leaching into a water supply through outdated or malfunctioning water supply infrastructure. Many inorganic contaminants, such as dissolved metals, can have toxic health effects. Feature 31, Inorganic Contaminants, is a precondition for all three project types that sets maximum safety limits for the presence of inorganic contaminants in drinking water.

Part 1: Dissolved Metals

Over 20 inorganic compounds might be present in a water supply. However, Part 1: Dissolved Metals of Feature 31 addresses the six most common and hazardous dissolved metal contaminants: lead, arsenic, antimony, mercury, nickel, and copper. **See Figure 6-3.** All water delivered to a project for human consumption (at least one water dispenser per project) must meet the following limits:

- lead—less than 0.01 mg/L
- arsenic—less than 0.01 mg/L
- antimony—less than 0.006 mg/L
- mercury—less than 0.002 mg/L
- nickel—less than 0.012 mg/L
- copper—less than 1.0 mg/L

Inorganic Contaminant Limits			
Inorganic Contaminant	**Limits***	**Contamination Sources**	**Potential Health Effects**
Lead	Less than 0.01	Enters environment from industry, mining, plumbing, gasoline, coal, and as a water additive	Affects red blood cell chemistry; delays normal physical and mental development in babies and young children; causes slight deficits in attention span, hearing, and learning in children
Arsenic	Less than 0.01	Natural processes; industrial activities and waste; pesticides; and smelting of copper, lead, and zinc ore	Acute and chronic toxicity; liver and kidney damage; decreases blood hemoglobin; carcinogenic
Antimony	Less than 0.006	Industrial production; municipal waste disposal; and manufacturing of flame retardants, ceramics, glass, batteries, fireworks, and explosives	Decreases longevity; alters blood levels of glucose and cholesterol in laboratory animals exposed at high levels over their lifetime
Mercury	Less than 0.002	Industrial waste, mining, pesticides, coal, eletrical equipment (batteries, lamps, and switches), smelting, and fossil fuel combustion	Acute and chronic toxicity; targets the kidneys and can cause nervous system disorders
Nickel	Less than 0.012	Occurs naturally in soils; groundwater, and surface water; used in electroplating, stainless steel and alloy products, mining, and refining	Allergic reaction; skin irritation; may cause respiratory issues if inhaled
Copper	Less than 1.0	Enters environment from metal plating, industrial and domestic waste, mining, and mineral leaching	Stomach and intestinal distress; liver and kidney damage; iron deficiency leading to anemia in high doses

* in mg/L

Figure 6-3. Six common and hazardous dissolved metals must be limited in a water supply to protect the health of building occupants.

BY THE NUMBERS

- *While there are environmental implications to the overreliance on bottled water, another concern is that the quality of bottled water is subject to degradation. In one study, levels of antimony in 48 brands of bottled water from 11 European countries increased by 90% after 6 months of storage due to antimony leaching from polyethylene terephthalate (PET) bottles, which are designated as recyclable "1."*

If a water supply is found to have levels of dissolved metals in excess of those limits, a filtration system can be used as a mitigation option. WELL lists two filtration options that are effective in reducing the levels of inorganic contaminants: reverse-osmosis filtration systems and kinetic degradation fluxion filters.

A *reverse-osmosis (RO) filtration system* is a water filtration system that uses a semipermeable membrane to filter water. Pressure forces untreated water against a very-fine, semipermeable membrane that primarily only allows water molecules to pass through. **See Figure 6-4.** The RO membranes are often used with several prefilter and postfilter components to minimize fouling of the membrane and remove chemicals that pass through the membrane.

RO filtration systems are effective at removing contaminants, but the filtration process wastes about 50% to 80% of the incoming supply water. Some RO filtration systems circulate this water back into the building supply to be used as graywater for nonpotable purposes. Also, large building-size systems can be costly.

A *kinetic degradation fluxion (KDF) filter* is a water filter that contains flakes or granules of a copper-and-zinc alloy. A KDF filter is highly effective (up to 98%) at removing many inorganic contaminants. However, KDF filters are not effective at removing organic chemicals or parasites.

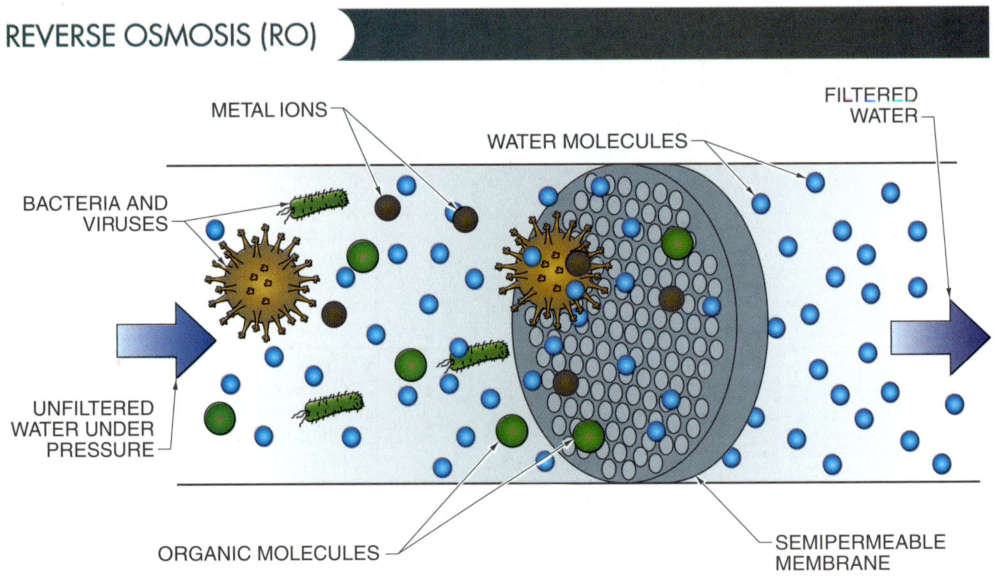

Figure 6-4. An RO filtration system uses a semipermeable membrane to separate water molecules from contaminants.

FEATURE 32, ORGANIC CONTAMINANTS

An *organic contaminant* is a human-made compound or chemical containing carbon atoms that has leached into ground and surface water from industrial activities, such as the production of plastics. While acute exposure to high levels of organic contaminants in a municipal water supply is rare, the cumulative effects of chronic exposure include severe damage to the nervous system and kidneys as well as an increased risk of cancer. Feature 32, Organic Contaminants, is a precondition for all three project types that lists maximum levels for eight organic contaminants.

Photo courtesy of USDA NRCS

Although surface water often already contains trace amounts of organic contaminants, industrial activities can leach higher levels of these contaminants into the water.

WELLNESS FACT

The three isomers of xylene (m, p, and o) all have the same chemical makeup—$C_6H_4(CH_3)_2$. The difference between the three is the placement of metyl groups, $(CH_3)_2$ on the C_6H_4 benzene ring.

Part 1: Organic Pollutants

Hundreds of types of organic compounds might find their way into a water supply. However, Part 1: Organic Pollutants of Feature 32 lists eight contaminants that can have a significant health effect on building occupants: styrene, benzene, ethylbenzene, polychlorinated biphenyls, vinyl chloride, toluene, xylenes (m, p, and o), and tetrachloroethylene. **See Figure 6-5.** All water delivered to a project for human consumption (at least one water dispenser per project) must meet the following limits:

- styrene—less than 0.0005 mg/L
- benzene—less than 0.001 mg/L
- ethylbenzene—less than 0.3 mg/L
- polychlorinated biphenyls (PCBs)—less than 0.0005 mg/L
- vinyl chloride—less than 0.002 mg/L
- toluene—less than 0.15 mg/L
- xylenes (m, p, and o)—less than 0.5 mg/L
- tetrachloroethylene—less than 0.005 mg/L

If a water supply is found to have levels of organic contaminants above those limits, a granular activated carbon filtration system can be used as a mitigation option. A *granular activated carbon (GAC) filtration system* is a water filtration system that uses oxygen-treated carbon to chemically bond with the organic contaminants in water. **See Figure 6-6.** GAC filtration systems are commonly used in homes as point-of-use filters for under-sink units, refrigerators with water dispensers, and additional faucets next to main drinking water faucets. GAC filtration systems are highly effective at removing organic contaminants, but they are ineffective at removing other substances, such as sodium and fluorine.

Proper monitoring and maintenance of a GAC filtration system is important because the pores of the carbon will fill up with the bonded organic compounds. A maintenance and filter-replacement schedule should be in place before final occupancy of the project.

Organic Contaminant Limits			
Organic Contaminant	**Limits***	**Contamination Sources**	**Potential Health Effects**
Styrene	Less than 0.0005	Discharge from rubber and plastic factories; leaching from landfills	Liver, kidney, or circulatory system problems
Benzene	Less than 0.001	Discharge from factories; leaching from gas storage tanks and landfills	Anemia; decrease in blood platelets; increased risk of cancer; bone marrow abnormalities
Ethylbenzene	Less than 0.3	Discharge from petroleum refineries	Liver or kidney problems
Polychlorinated biphenyls (PCBs)	Less than 0.0005	Runoff from landfills; discharge of waste chemicals	Skin changes; thymus gland problems; immune deficiencies; reproductive or nervous system difficulties; increased risk of cancer
Vinyl chloride	Less than 0.002	Leaching from PVC pipes; discharge from plastic factories	Increased risk of cancer
Toluene	Less than 0.15	Discharge from petroleum factories	Nervous system, kidney, or liver problems
Xylenes (total: m, p, and o)	Less than 0.5	Discharge from petroleum factories; discharge from chemical factories	Nervous system damage
Tetrachloroethylene	Less than 0.005	Discharge from factories and dry cleaners	Liver problems; increased risk of cancer

* in mg/L

Figure 6-5. Organic contaminants in a water supply can pose a significant health hazard if ingested.

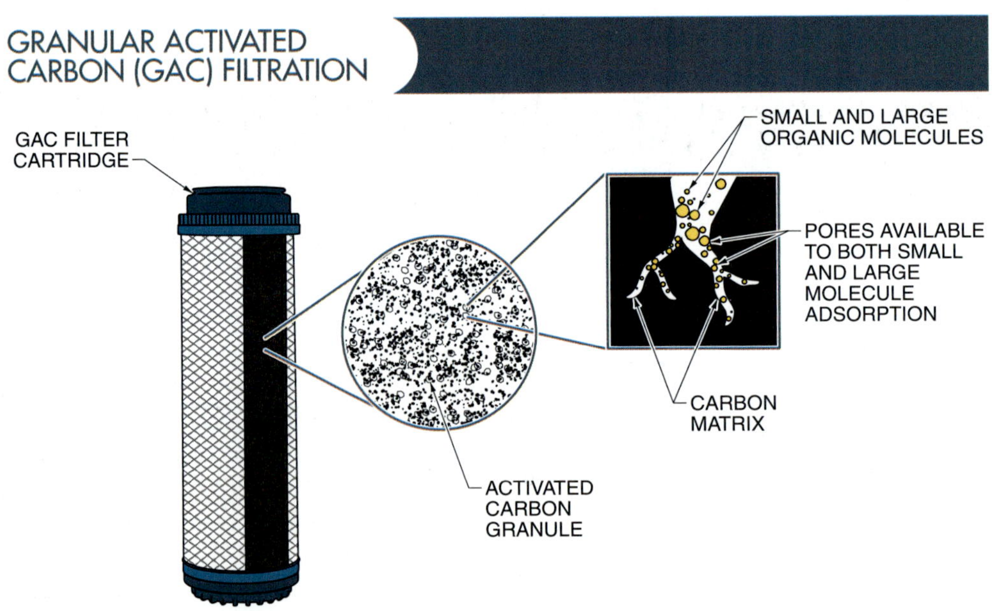

Figure 6-6. GAC chemically bonds with organic contaminants, removing them from a water supply.

FEATURE 33, AGRICULTURAL CONTAMINANTS

An *agricultural contaminant* is a chemical pesticide, herbicide, or fertilizer that can be harmful to humans, animals, or the environment if it leaches into a water supply. These chemicals can mix with water and seep into the soil and groundwater, or they may be absorbed by the plants they are meant to help. From there, these chemicals may be introduced into streams, lakes, reservoirs, and other sources of drinking water and consumed by animals or humans. Feature 33, Agricultural Contaminants, is a precondition for all three project types that sets limits for herbicides, pesticides, and fertilizers. **See Figure 6-7.**

BY THE NUMBERS

- *In the 1990s, a U.S. Geological Survey detected pesticide compounds in virtually every stream in agricultural, urban, and mixed-use areas, as well as in 30% to 60% of the groundwater.*

Pesticides or fertilizers applied aerially can drift into areas where they are not intended.

Part 1: Herbicides and Pesticides

A *pesticide* is a chemical that is used to destroy, repel, or control plants or animals. Pesticides can be harmful to human or animal health. Examples of pesticides include herbicides, insecticides, fungicides, and rodenticides.

An *herbicide* is a type of pesticide that contains chemicals used to destroy or inhibit the growth of unwanted plants. Herbicides are used in farming and for the control of invasive plants or weeds. They may be applied to the soil before a crop is planted, or they may be applied to the plant once it emerges from the soil.

Agricultural Contaminant Limits		
Agricultural Contaminant	Limits*	Potential Health Effects
Atrazine	Less than 0.001	Cardiovascular system or reproductive problems
Simazine	Less than 0.002	Damage to liver and thyroid; reproductive difficulties
Glyphosate	Less than 0.70	Kidney problems; reproductive difficulties
2,4-dichlorophenoxyacetic acid	Less than 0.07	Gastroentric distress; mild central nervous system depression; liver and kidney damage
Nitrate	Less than 50 (10 as nitrogen)	Methemoglobinemia in infants or high-risk patients

* in mg/L

Figure 6-7. Agricultural contaminants pose a health hazard due to contaminant-containing runoff and the absorption of pesticides, herbicides, and fertilizers by the ground or plants.

Part 1: Herbicides and Pesticides of Feature 33 requires that all water being delivered to the project area for human consumption (at least one water dispenser per project) be tested to ensure that the levels of pesticides and herbicides do not exceed maximum safety limits. If herbicide and pesticide levels exceed the maximum safety levels, a GAC filtration system can be used to effectively remove these chemicals from the water supply. Limits are set for the following four herbicides and pesticides:

- Atrazine levels must be less than 0.001 mg/L. *Atrazine* is a pesticide that is among the most widely used pesticides in the United States to control broadleaf weeds in crops. Atrazine is also among the most commonly detected pesticides in drinking water.
- Simazine levels must be less than 0.002 mg/L. *Simazine* is a popular herbicide that is used to control weeds. High levels of simazine exposure over a short period can cause weight loss and blood damage.
- Glyphosate levels must be less than 0.70 mg/L. *Glyphosate* is a nonselective herbicide that is used in many pesticide formulations, which may result in human exposure through its normal use due to spray drift, residues on food crops, and runoff into drinking water sources.
- 2,4 dichlorophenoxyacetic acid levels must be less than 0.07 mg/L. *2,4-dichlorophenoxyacetic acid (2,4-D)* is a major herbicide that is likely to run off or leach into ground and surface water sources.

Part 2: Fertilizers

A *fertilizer* is a compound that contains nutrients that encourage the growth of a plant. A key component of many fertilizers is nitrogen in the form of ammonium nitrate ($N_2H_4O_3$), ammonium sulfate ($[NH_4]_2SO_4$), or urea (CH_4N_2O). These forms of nitrogen are then converted to nitrite (NO_2) and to nitrate (NO_3) by bacteria in the soil. Unused nitrates in the soil can find their way into groundwater. In addition to fertilizers, nitrates can be introduced into the ecosystem through livestock waste and septic tank effluent.

Part 2: Fertilizers of Feature 33 specifies that the nitrate level for all water being delivered to the project area for human consumption (at least one water dispenser per project) must contain less than 50 mg/L of nitrates (10 mg/L as nitrogen). If this level is exceeded, measures must be taken to lower the level. Excessive nitrate exposure can cause methemoglobinemia, a serious blood disorder in which hemoglobin in the blood carries but is unable to release oxygen effectively to the body tissue. The use of a GAC filtration system is the most effective way of reducing nitrate levels in a water supply.

FEATURE 34, PUBLIC WATER ADDITIVES

In order to make public water supplies safe for consumption, chemicals are often added at a water treatment plant to help prevent the spread of disease and improve water quality. Since the early 1900s, the main chemicals added to public water supplies have been chlorine and chloramine disinfectants. In addition, some municipal water utilities add fluoride to the water to help prevent tooth decay. While beneficial to public health in small amounts, excessive amounts of these chemicals and their by-products can be detrimental to human health. Feature 34, Public Water Additives, is a precondition for all three project types that requires disinfectant, disinfectant by-product, and fluoride levels to be under set limits. **See Figure 6-8.**

Part 1: Disinfectants

A *disinfectant* is a chemical that is used to control or destroy harmful microorganisms as well as prevent their formation on inanimate objects and surfaces or in liquids. Part 1: Disinfectants of Feature 34 requires that the amount of chlorine and chloramine (two disinfectants) in a water supply for human consumption and showers/baths be limited to less than 0.6 mg/L and 4 mg/L, respectively.

Chlorine is a highly irritating, greenish-yellow gaseous halogen that can be introduced into a water supply as a gas, sodium hypochlorite solution, or calcium hypochlorite solid. Chlorine is inexpensive and highly effective at killing microorganisms in a water supply. It can, however, cause health problems because it can react with other organic matter in the water to produce unwanted compounds. A GAC filtration system can be used to lower the chlorine level if measurements indicate that it exceeds the limits set in Part 1.

Chloramine is a disinfectant formed when ammonia is added to chlorine and is commonly used as a secondary disinfectant in public water systems. While chloramine is not as effective at killing microorganisms as chlorine, it does not dissipate into the air as quickly, which allows it more time to disinfect. Chloramine also generates fewer by-products. Excess levels of chloramine may require the use of a GAC filtration system in order to be effectively lowered.

Part 2: Disinfectant By-products

A *disinfectant by-product (DBP)* is a compound that forms when chlorine and, to a slightly lesser extent, chloramine react with organic materials in a water supply. Part 2: Disinfectant By-products of Feature 34 requires that the amount of two groups of DBPs, trihalomethanes and haloacetic acids, in a water supply be limited to less than 0.08 mg/L and 0.06 mg/L, respectively. A *trihalomethane (THM)* is a disinfectant by-product that is formed when chlorine reacts with organic matter in water. A *haloacetic acid (HAA)* is a disinfectant by-product formed when chlorine or chloramine reacts with organic matter in water. Both THMs and HAAs may cause skin and eye irritation and also increase the risk of cancer. GAC filtration systems are highly effective at removing THMs and HAAs from a water supply.

Part 3: Fluoride

Fluoride is a naturally occurring chemical that prevents or helps reverse tooth decay. Although natural fluoride may be present in ground and surface water, many municipal water utilities add fluoride to drinking water. Fluoride has been instrumental in lowering the rate of tooth decay in the general population over the last 70 years. However, the fluoride added to water combined with the fluoride found in toothpastes, mouthwashes, and other dental products may lead to excessive fluoride consumption. **See Figure 6-9.**

Public Water Additive Limits		
Additive	**Limits***	**Potential Health Effects**
Chlorine	Less than 0.6	Eye/nose irritation; stomach discomfort
Chloramine	Less than 4	Eye/nose irritation; stomach discomfort; anemia
Trihalomethanes	Less than 0.08	Bladder cancer; heart, lungs, kidney, liver, and central nervous system damage
Haloacetic acids	Less than 0.06	Lung, liver, and kidney problems
Fluoride	Less than 4.0	Dental fluorosis (spotting or pitting on teeth)

* in mg/L

Figure 6-8. While beneficial in small amounts, excessive amounts of the chemicals from public water additives and their by-products can be detrimental to human health.

FLUORIDE SOURCES

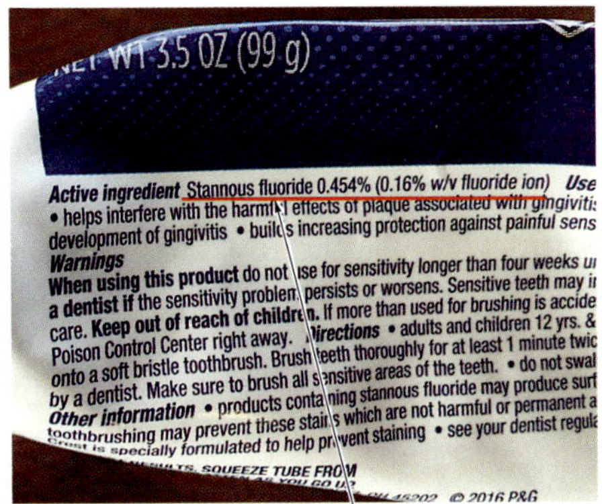

Figure 6-9. In addition to most municipal drinking water, fluoride can be found in a large number of oral-care products.

Excessive fluoride consumption can cause dental fluorosis. Dental fluorosis is a condition in which white spots can form on a tooth surface for mild cases. Excessive enamel damage such as pitting may occur with highly excessive consumption of fluoride. Part 3: Fluoride of Feature 34 requires that the amount of fluoride in a water supply be limited to less than 4.0 mg/L. Although excessive fluoride can be difficult to remove from drinking water, reverse osmosis or distillation can be effective.

FEATURE 35, PERIODIC WATER QUALITY TESTING

The level of inorganic metals in an incoming water supply may vary due to changes in temperature, pH levels, and the weather. When incoming water is warm or has a low pH (more acidic), it may leach metals from storage containers or supply piping and fixtures at a high rate. In order to maintain safe levels of certain inorganic metals, it is important to routinely test incoming water for changes. Feature 35, Periodic Water Quality Testing, is an optimization for the New and Existing Interiors and New and Existing Buildings project types that requires quarterly testing as well as detailed records, responses, and remediation plans.

Part 1: Quarterly Testing

Part 1: Quarterly Testing of Feature 35 requires tests for inorganic metals once every quarter. The tests must measure the levels of lead, arsenic, mercury, and copper in the water supply. A procedure must be in place to document and report the quarterly tests to IWBI once a year. The limits for dissolved metals are listed in Feature 31, Inorganic Contaminants.

Part 2: Water Data Record Keeping and Response

Part 2: Water Data Record Keeping and Response of Feature 35 requires that a policy be written concerning how to enforce water quality monitoring and record-keeping strategies. The policy must specify that detailed records of testing and inspections be kept for a minimum of three years. The policy must also include a detailed plan for taking action if the levels of dissolved metals exceed the limits set in Feature 31, Inorganic Contaminants.

Continued testing and reporting ensures that a project maintains water quality and safety. It may also verify that the water remediation systems put in place are working at the designed levels or may indicate system deficiencies. For example, a project's water filtration system operations and maintenance (O&M) plan may call for the filters to be changed once a year. As the year progresses, the levels of inorganic contaminants steadily rise and approach critical levels toward the end of the year. However, the levels return to an acceptable range after the yearly service and maintenance is completed. In this example, testing can indicate that the water filtration systems require service and maintenance more frequently than originally designated.

FEATURE 36, WATER TREATMENT

Once an incoming water supply has met the required preconditions for the Water Concept, water quality and safety should be maintained. Feature 36, Water Treatment, is an optimization for all three project types that includes five parts of requirements. The first three parts address water filtration and sanitation. These three parts involve treating all water that is being delivered to the project area for human consumption. The fourth part addresses the operation, maintenance, and record keeping of the filtration/sanitation systems. The fifth part addresses a Legionella control plan.

Part 1: Organic Chemical Removal

Part 1: Organic Chemical Removal of Feature 36 requires the installation of activated carbon filters, such as GAC filters, to remove organic chemicals from any water that is used for human consumption.

Part 2: Sediment Filter

Part 2: Sediment Filter of Feature 36 requires the use of a sediment filter rated to remove suspended solids from any water that is used for human consumption. The pore size of the filter must be 1.5 µm or less. Glass-microfiber filters or RO filtration systems can be used to remove suspended solids from a water supply.

Part 3: Microbial Elimination

Part 3: Microbial Elimination of Feature 36 lists two methods of microbial elimination for any water that is used for human consumption: ultraviolet germicidal irradiation filters and filters rated by the National Science Foundation (NSF) to remove microbial cysts. *Ultraviolet germicidal irradiation (UVGI)* is a sterilization method that uses UV light to break down microorganisms by destroying their DNA. UVGI is used in a variety of applications, such as food, air, and water purification.

UVGI systems have been used in medical sanitation and sterile work facilities for many years. The NSF rates filters according to their ability to remove microbial cysts, such as Cryptosporidium and Giardia, which can cause intestinal illness. A *microbial cyst* is a microorganism in its dormant state that is resistant to typical disinfection methods.

Part 4: Water Quality Maintenance

Part 4: Water Quality Maintenance of Feature 36 requires verification that the filtration/sanitation systems used for a building continue to operate as designed. Every year the IWBI must be provided evidence that the systems are properly maintained and serviced per manufacturer recommendations. Records must be kept for a minimum of three years.

Part 5: Legionella Control

Legionella is a bacterium that is found in freshwater and can cause a serious form of pneumonia called Legionnaires' disease. Legionella is easily spread through the air in mist or water vapor. The bacteria can then enter the lungs, which poses a serious issue for people who are high-risk for lung issues or have weakened immune systems. Sources of Legionella include hot-water tanks, cooling towers, large plumbing systems, dental water lines, and aesthetic water features. Part 5: Legionella Control of Feature 36 requires a comprehensive narrative that addresses the control of Legionella in a building. **See Figure 6-10.**

BY THE NUMBERS

- *The Institute of Medicine (IOM) recommends that women consume approximately 2.7 L (91 oz) of water per day and that men consume 3.7 L (125 oz) of water per day from all sources. Active people or people in warmer environments may require larger amounts of water.*

LEGIONELLA CONTROL NARRATIVE

- Formation of Legionella management team
- Water system inventory and production of process flow diagrams
- Hazard analysis of water assets
- Identification of critical control points
- Maintenance and control measures, monitoring, establishment of performance limits, and corrective actions
- Documentation, verification, and validation procedures

Figure 6-10. A Legionella control narrative is a point-by-point document for how to control exposure to Legionella.

FEATURE 37, DRINKING WATER PROMOTION

Adequate hydration is necessary for the proper physiological functions of the human body. Mild dehydration can result in muscle cramps, dry skin, and headaches. Severe dehydration is a life-threatening condition that can cause confusion, rapid heartbeat and breathing, shock, and delirium.

Even though WELL requires that water for human consumption be safe, none of the preconditions in the Water Concept address the taste or appearance of water. However, these qualities are addressed in Feature 37, Drinking Water Promotion, which is an optimization that addresses the properties of drinking water taste, the location and number of drinking water dispensers, and the maintenance of drinking water equipment. The first part of this feature applies to all three project types, but the second and third parts only apply to the New and Existing Interiors and New and Existing Buildings types.

Part 1: Drinking Water Taste Properties

Part 1: Drinking Water Taste Properties of Feature 37 states the maximum levels of dissolved minerals and total dissolved solids that the drinking water delivered to a project area may contain. The minerals that may affect the properties of drinking water taste include aluminum, chloride, manganese, sodium, sulfate, iron, and zinc. **See Figure 6-11.** All water delivered to a project must meet the following limits:

- aluminum—less than 0.2 mg/L
- chloride—less than 250 mg/L
- manganese—less than 0.05 mg/L
- sodium—less than 270 mg/L
- sulfate—less than 250 mg/L
- iron—less than 0.3 mg/L
- zinc—less than 5 mg/L
- total dissolved solids—less than 500 mg/L

Additional filtration is required if any level of the dissolved minerals or total dissolved solids exceeds the maximum set in the standard. The additional filtration may be installed for the whole system or the point of use. **See Figure 6-12.**

Part 2: Drinking Water Access

Quality drinking water must be readily available for building occupants and guests to use. If they have to walk for several hundred meters or go to another floor to obtain water, there is less chance of them drinking water throughout the day. Part 2: Drinking Water Access of Feature 37 requires at least one drinking water dispenser per floor, no farther than 30 m (100′) from all parts of the regularly occupied floor space. **See Figure 6-13.** All dispensers must meet the drinking water taste properties of Part 1.

Drinking Water Taste Properties		
Dissolved Mineral	**Limits***	**Undesired Effects**
Aluminum	Less than 0.2	Color
Chloride	Less than 250	Salty taste
Manganese	Less than 0.05	Color
Sodium	Less than 270	Salty taste
Sulfate	Less than 250	Salty taste
Iron	Less than 0.3	Rusty color; metallic taste; odor
Zinc	Less than 5	Metallic taste
Total dissolved solids (TDS)	Less than 500	Hardness; color; salty taste; odor

* in mg/L

Figure 6-11. Dissolved minerals in drinking water may give the water an undesirable taste, color, or odor.

DRINKING WATER FIXTURE FILTERS

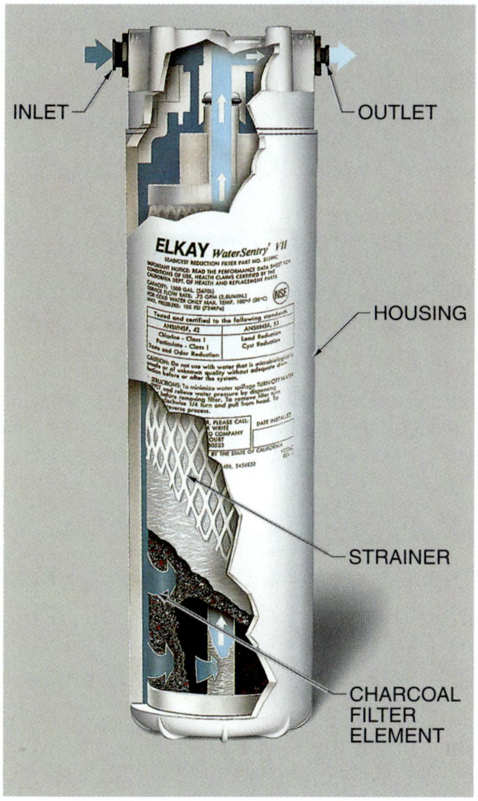

Figure 6-12. A point-of-use filter, such as GAC filters used in water coolers or drinking fountains, can effectively remove dissolved minerals from drinking water.

DRINKING WATER ACCESS

Figure 6-13. Easily accessible drinking water dispensers on every floor provide regular building occupants with quality drinking water.

Part 3: Water Dispenser Maintenance

Part 3: Water Dispenser Maintenance of Feature 37 requires that drinking water dispensers be properly serviced and maintained to ensure that they are safe, are working properly, and continue to deliver high-quality water. Part 3a requires daily cleaning of mouthpieces, guards, and basins to prevent lime and calcium buildup. Part 3b requires quarterly cleaning of outlet screens and aerators to remove debris and sediment.

KEY TERMS AND DEFINITIONS

agricultural contaminant: A chemical pesticide, herbicide, or fertilizer that can be harmful to humans, animals, or the environment if it leaches into a water supply.

atrazine: A pesticide that is among the most widely used pesticides in the United States to control broadleaf weeds in crops.

chloramine: A disinfectant formed when ammonia is added to chlorine and is commonly used as a secondary disinfectant in public water systems.

chlorine: A highly irritating, greenish-yellow gaseous halogen that can be introduced into a water supply as a gas, sodium hypochlorite solution, or calcium hypochlorite solid.

coliform: A microorganism that includes bacteria such as E. coli.

disinfectant: A chemical that is used to control or destroy harmful microorganisms as well as prevent their formation on inanimate objects and surfaces or in liquids.

disinfectant by-product (DBP): A compound that forms when chlorine and, to a slightly lesser extent, chloramine react with organic materials in a water supply.

fertilizer: A compound that contains nutrients that encourage the growth of a plant.

fluoride: A naturally occurring chemical that prevents or helps reverse tooth decay.

glyphosate: A nonselective herbicide that is used in many pesticide formulations, which may result in human exposure through its normal use due to spray drift, residues on food crops, and runoff into drinking water sources.

granular activated carbon (GAC) filtration system: A water filtration system that uses oxygen-treated carbon to chemically bond with the organic contaminants in water.

haloacetic acid (HAA): A disinfectant by-product formed when chlorine or chloramine reacts with organic matter in water.

herbicide: A type of pesticide that contains chemicals used to destroy or inhibit the growth of unwanted plants.

inorganic contaminant: An element or compound that may be found in a water supply, occurring from natural sources such as the geology of a location, resulting from human activities such as mining and industry, or leaching into a water supply through outdated or malfunctioning water supply infrastructure.

kinetic degradation fluxion (KDF) filter: A water filter that contains the flakes or granules of a copper-and-zinc alloy.

Legionella: A bacterium that is found in freshwater and can cause a serious form of pneumonia called Legionnaires' disease.

microbial cyst: A microorganism in its dormant state that is resistant to typical disinfection methods.

nephelometric turbidity unit (NTU): The unit of measure for the turbidity of water.

nonpotable water: Water that is not fit for human consumption.

organic contaminant: A human-made compound or chemical containing carbon atoms that has leached into ground and surface water from industrial activities, such as the production of plastics.

pathogen: An infectious biological agent such as a bacterium, virus, or fungus that is capable of causing disease in its host.

pesticide: A chemical that is used to destroy, repel, or control plants or animals.

KEY TERMS AND DEFINITIONS *(continued)*

potable water: Water that is fit for human consumption.

process water: Water that is used for cooling towers, boilers, and industrial processes.

reverse-osmosis (RO) filtration system: A water filtration system that uses a semipermeable membrane to filter water.

simazine: A popular herbicide that is used to control weeds.

trihalomethane (THM): A disinfectant by-product that is formed when chlorine reacts with organic matter in water.

turbidity: The amount of cloudiness in a liquid caused by suspended solids that are invisible to the naked eye.

2,4-dichlorophenoxyacetic acid (2,4-D): A major herbicide that is likely to run off or leach into ground and surface water sources.

ultraviolet germicidal irradiation (UVGI): A sterilization method that uses UV light to break down microorganisms by destroying their DNA.

CHAPTER REVIEW

Completion

______________ **1.** ___ is the amount of cloudiness in a liquid caused by suspended solids that are invisible to the naked eye.

______________ **2.** The ___ the turbidity, the greater the amount of sedimentation or other substances that the water sample may contain.

______________ **3.** Part 1: Sediment of Feature 30, Fundamental Water Quality, requires that the turbidity of a water sample be less than ___ NTU.

______________ **4.** Part 2: Microorganisms of Feature 30, Fundamental Water Quality, requires that ___ not be present in a water sample.

______________ **5.** A(n) ___ is an infectious biological agent such as a bacterium, virus, or fungus that is capable of causing disease in its host.

______________ **6.** A reverse-osmosis (RO) or ___ filtration system is effective in reducing levels of inorganic contaminants.

______________ **7.** A(n) ___ contaminant is a human-made compound or chemical containing carbon atoms that has leached into ground or surface water from industrial activities, such as the production of plastics.

______________ **8.** ___ filter systems are highly effective in removing organic contaminants, but they are ineffective at removing other substances, such as sodium and fluorine.

______________ **9.** A(n) ___ contaminant is a chemical pesticide, herbicide, or fertilizer that can be harmful to humans, animals, or the environment if it leaches into a water supply.

CHAPTER REVIEW *(continued)*

______________ **10.** A(n) ___ is a compound that contains nutrients that encourage the growth of a plant.

______________ **11.** Part 2: Fertilizers of Feature 33, Agricultural Contaminants, specifies that the nitrate level for all water being delivered to a project area for human consumption must be less than ___ mg/L.

______________ **12.** A(n) ___ is a chemical that is used to control or destroy harmful microorganisms as well as prevent their formation on inanimate objects and surfaces or in liquids.

______________ **13.** A(n) ___ is a compound that forms when chlorine and, to a slightly lesser extent, chloramine react with organic materials in a water supply.

______________ **14.** ___ is a naturally occurring chemical that prevents or helps reverse tooth decay.

______________ **15.** Any water for a project area that is meant for human consumption must contain less than ___ mg/L of fluoride.

______________ **16.** Part 1: Quarterly Testing of Feature 35, Periodic Water Quality Testing, requires that a procedure be in place to document and report ___ water testing, on a yearly basis, to IWBI.

______________ **17.** A(n) ___, such as a GAC filter, should be used to treat water for human consumption per Part 1: Organic Chemical Removal of Feature 36, Water Treatment.

______________ **18.** The sediment filter described in Part 2: Sediment Filter of Feature 36, Water Treatment, must be rated to remove ___ from any water for human consumption.

______________ **19.** For Part 4: Water Quality Maintenance of Feature 36, Water Treatment, record keeping is required for a minimum of ___ years.

______________ **20.** Sources of ___ bacterium include hot-water tanks, cooling towers, large plumbing systems, dental water lines, and aesthetic water features.

______________ **21.** Part 1: Drinking Water Taste Properties of Feature 37, Drinking Water Promotion, states that the maximum levels of total dissolved solids in the drinking water delivered to a project area must be less than ___ mg/L.

______________ **22.** Part 2: Drinking Water Access of Feature 37, Drinking Water Promotion, requires at least ___ drinking water dispenser(s) per floor, no farther than 30 m (100′) from all parts of the regularly occupied floor area.

Short Answer

1. List the five preconditions in the Water Concept.

2. List the three optimizations in the Water Concept.

CHAPTER REVIEW *(continued)*

3. List the six dissolved metals and their respective limits for Feature 31, Inorganic Contaminants.

4. List the contaminants and their respective limits for Feature 32, Organic Contaminants.

5. List the herbicides and pesticides and their limits that are addressed in Part 1: Herbicides and Pesticides of Feature 33, Agricultural Contaminants.

6. What are the two disinfectants addressed in Part 1: Disinfectants of Feature 34, Public Water Additives, their limits, and their differences?

7. List the two disinfectant by-products and their limits addressed in Part 2: Disinfectant By-products of Feature 34, Public Water Additives.

8. What is the main health hazard and the effects of excessive amounts of fluoride in drinking water?

9. List the four substances that are required to be tested per Part 1: Quarterly Testing of Feature 35, Periodic Water Quality Testing.

10. What are the three requirements for the written policy addressed by Part 2: Water Data Record Keeping and Response of Feature 35, Periodic Water Quality Testing?

11. Part 3: Microbial Elimination of Feature 36, Water Treatment, requires one of which two strategies?

12. List the six steps for Part 5: Legionella Control of Feature 36, Water Treatment.

CHAPTER REVIEW *(continued)*

13. What are the symptoms of mild and severe dehydration?

14. List the limitations for the seven dissolved minerals and total dissolved solids in a water supply per Part 1: Drinking Water Taste Properties of Feature 37, Drinking Water Promotion.

15. What are the daily and quarterly cleaning requirements for Part 3: Water Dispenser Maintenance of Feature 37, Drinking Water Promotion?

WELL AP EXAM PRACTICE QUESTIONS

1. The test of a project's incoming water supply reveals nickel levels at 0.1 mg/L and lead levels at 0.05 mg/L. Which of the following is the best strategy for lowering these levels to meet the requirements of Feature 31, Inorganic Contaminants?
   A. Granular activated carbon (GAC) filter
   B. Sediment filter
   C. Ultraviolet germicidal irradiation (UVGI) filter
   D. Reverse-osmosis (RO) filtration system

2. What is required by Part 5: Legionella Control of Feature 36, Water Treatment?
   A. A professional narrative that addresses the control of Legionella in the building
   B. The installation of UVGI filters
   C. Monthly testing of areas that might be at risk of Legionella contamination
   D. The elimination of any water features that might pose a risk of Legionella contamination

3. Nitrate levels in excess of 50 mg/L in the water that is delivered to a project area for human consumption can cause which of the following major health risks?
   A. Melanoma carcinoma
   B. Methemoglobinemia
   C. Legionnaires' disease
   D. Dental fluorosis

4. Which of the following features requires turbidity of less than 1.0 NTU and no coliforms detected in an incoming water supply?
   A. Feature 30, Fundamental Water Quality
   B. Feature 32, Organic Contaminants
   C. Feature 34, Public Water Additives
   D. Feature 35, Periodic Water Quality Testing

5. Which feature in the Water Concept focuses on improving the taste and appearance of water in order to increase occupant hydration?
   A. Feature 30, Fundamental Water Quality
   B. Feature 33, Agricultural Contaminants
   C. Feature 35, Periodic Water Quality Testing
   D. Feature 37, Drinking Water Promotion

CHAPTER SEVEN

NOURISHMENT

Adequate, healthy nourishment through proper nutrition is necessary for human well-being. However, poor nutrition education and practices have lead to a global obesity crisis, as well as an increase in other diet-related diseases. Left unchecked, these problems will continue to grow and affect millions of more people every year. The primary goals of the Nourishment Concept in the WELL Building Standard are to increase access to healthy foods and beverages, help people make informed dietary choices, and reduce the possibility of exposure to harmful pathogens and allergens in food.

OBJECTIVES

- Identify the features in the Nourishment Concept as preconditions or optimizations.
- Explain how a human body uses the six essential nutrients.
- Explain the dietary factors that lead to various nourishment-related diseases and their implications on human health.
- Explain the roles of body mass index (BMI), nutritional information, and serving sizes in healthy eating.
- Identify which features in the Nourishment Concept ban unhealthy foods and which features promote healthy foods and eating behaviors.
- Explain how a culture of healthy eating behaviors is encouraged through the features in the Nourishment Concept.
- List the eight food allergies and one food sensitivity addressed in the Nourishment Concept.
- Explain the safe food practices encouraged by the features in the Nourishment Concept.
- Describe special diets and how to address them.
- Describe the on-site spaces in which food can be produced.
- Describe the benefits of mindful eating and the requirements for mindful eating spaces.
- Explain how to apply specific features in the Nourishment Concept based on different project situations.

KEY TERMS

- body mass index (BMI)
- cancer
- carbohydrate
- cardiovascular disease
- diabetes
- gluten
- humanely raised food
- hypertension
- lipid
- macronutrient
- malnutrition
- mineral
- nutrient
- obesity
- organic food
- processed food
- protein
- trans fat
- vitamin

Learner Resources

ATPeResources.com/QuickLinks
Access Code: 935534

NOURISHMENT CONCEPT

Nutrition is a central principle of nourishment because it helps maintain human health, manage body weight, and prevent chronic diseases associated with obesity. Many factors influence the level of nutrition people achieve, including the places where people live, work, and choose to eat. WELL aims to promote healthy and informed dietary choices within the built environment.

The Nourishment Concept in WELL includes 15 features that promote strategies to increase the availability of healthy foods and encourage better nutrition and eating habits. Core and Shell projects must meet two preconditions, but five optimizations can be pursued for higher certification. Both New and Existing Interiors and New and Existing Buildings must meet eight preconditions, but seven optimizations can be pursued for higher certification. **See Figure 7-1.**

Sullivan University

Healthier and more nutritious food choices should be promoted to improve the health and well-being of building occupants.

NUTRITION BASICS

In order to understand the requirements of many features in the Nourishment Concept, it is necessary to understand the basics of nutrition. A *nutrient* is a chemical that is required for metabolic processes, which must be taken from food or another external source. Nutrients are important to the aims of the Nourishment Concept because they are responsible for providing energy and helping the body grow, maintain, and repair its bones, muscles, organs, and cells. *Malnutrition* is a condition that results from insufficient nutrient intake, excess nutrient intake, or nutrient intake in the wrong proportions.

BY THE NUMBERS

- *In the United States, only 8% of people consume the recommended 4 servings of fruit per day and only 6% consume the recommended 5 servings of vegetables per day.*
- *The World Health Organization (WHO) reports that 2.7 million deaths worldwide are attributed to insufficient fruit and vegetable intake, which is why several features in the Nourishment Concept focus on the availability and consumption of fruits and vegetables.*

The body requires six essential nutrients—proteins, carbohydrates, lipids, water, vitamins, and minerals. These six essential nutrients are categorized as macronutrients or micronutrients.

Macronutrients

A *macronutrient* is a nutrient needed by the body in large amounts. Macronutrients include proteins, carbohydrates, lipids, and water. Proteins, carbohydrates, and lipids are needed in large amounts to supply energy for the body. Although water does not supply energy, it is considered a macronutrient because the body requires a large amount (approximately 2.7 L to 3.7 L daily) to function properly.

Proteins. A *protein* is an energy-providing nutrient that is made of carbon, hydrogen, oxygen, and nitrogen assembled in chains of amino acids. Some of the key functions of protein include building muscle, providing cellular structure for organs and tissues, transporting molecules, and supplying energy.

WELL Building Standard Features: Nourishment Concept...						
Features	Project Type			Verification Documentation		
	Core and Shell	New and Existing Interiors	New and Existing Buildings	Letter of Assurance	Annotated Documents	On-Site Assessment
38, Fruits and Vegetables						
Part 1: Fruit and Vegetable Variety	–	P	P		Operations Schedule	Spot Check
Part 2: Fruit and Vegetable Promotion	–	P	P		Operations Schedule	Spot Check
39, Processed Foods						
Part 1: Refined Ingredient Restrictions	P	P	P		Operations Schedule	Spot Check
Part 2: Trans Fat Ban	P	P	P		Operations Schedule	Spot Check
40, Food Allergies						
Part 1: Food Allergy Labeling	P	P	P		Operations Schedule	Spot Check
41, Hand Washing						
Part 1: Hand Washing Supplies	–	P	P		Operations Schedule	Spot Check
Part 2: Contamination Reduction	–	P	P			Visual Inspection
Part 3: Sink Dimensions	–	P	P	Architect		Spot Check
42, Food Contamination						
Part 1: Cold Storage	–	P	P	Owner		Spot Check
43, Artificial Ingredients						
Part 1: Artificial Substance Labeling	O	P	P		Operations Schedule	Spot Check
44, Nutritional Information						
Part 1: Detailed Nutritional Information	O	P	P			Visual Inspection
45, Food Advertising						
Part 1: Advertising and Environmental Cues	O	P	P			Visual Inspection
Part 2: Nutritional Messaging	O	P	P			Visual Inspection
46, Safe Food Preparation Materials						
Part 1: Cooking Material	–	O	O		Operations Schedule	Spot Check
Part 2: Cutting Surfaces	–	O	O		Operations Schedule	Spot Check
47, Serving Sizes						
Part 1: Meal Sizes	–	O	O		Operations Schedule	Spot Check
Part 2: Dishware	–	O	O		Operations Schedule	Spot Check
48, Special Diets						
Part 1: Food Alternatives	–	O	O		Operations Schedule	

Figure 7-1. (*continued on next page*)

...WELL Building Standard Features: Nourishment Concept						
Features	Project Type			Verification Documentation		
	Core and Shell	New and Existing Interiors	New and Existing Buildings	Letter of Assurance	Annotated Documents	On-Site Assessment
49, Responsible Food Production						
Part 1: Sustainable Agriculture	–	O	O		Operations Schedule	
Part 2: Humane Agriculture	–	O	O		Operations Schedule	
50, Food Storage						
Part 1: Storage Capacity	–	O	O	Owner		Spot Check
51, Food Production						
Part 1: Gardening Space	O	O	O	Owner		Spot Check
Part 2: Planting Support	O	O	O	Owner		Spot Check
52, Mindful Eating						
Part 1: Eating Spaces	O	O	O		Architectural Drawing	
Part 2: Break Area Furnishings	O	O	O	Owner		Spot Check

Figure 7-1. The Nourishment Concept in WELL includes 15 features that promote strategies to increase the availability of healthy foods and encourage better nutrition and eating habits.

Carbohydrates. A *carbohydrate* is any of a group of organic compounds that includes sugars, starches, celluloses, and gums and serves as a major energy source to support bodily functions and physical activity. The body breaks down sugar and starch into glucose to be used as energy. A high level of glucose in the blood may be linked to diabetes.

BY THE NUMBERS

- *Sugar often accounts for more than 500 calories in a person's daily diet, contributing greatly to the global obesity pandemic.*

Lipids. A *lipid* is an energy-providing nutrient made from fatty acids. Lipids include solid fats and oils. Lipids come from both animal and plant sources, and they are also produced naturally within the human body. Lipids from animal sources are known as saturated fats and often referred to as solid fats because they are solid at room temperature. Lipids from plant sources are known as unsaturated fats and often referred to as liquid fats (oils) because they are liquid at room temperature. Unsaturated fats are considered a healthy form of lipids.

Water. Water is believed to be the most important nutrient. Water is necessary for just about every process that occurs within the body, including transporting nutrients, carrying away waste, providing hydration, and normalizing body temperature.

Micronutrients

A micronutrient is a nutrient needed by the body in small amounts. Micronutrients include vitamins and minerals. Micronutrients do not supply energy to the body, but they do play an essential role in maintaining and regulating body functions.

Vitamins. A *vitamin* is an organic nutrient that is required in small amounts to help regulate body processes. Common vitamins include vitamins A, B_6, B_{12}, C, D, E, and K, which can all be found in foods.

Minerals. A *mineral* is an inorganic nutrient that is required in small amounts to help regulate body processes. The major minerals include calcium, phosphorus, magnesium, potassium, sodium, and chloride.

BY THE NUMBERS

- *According to calculations using the body mass index (BMI), more than two-thirds of adults in the United States are overweight, and more than a third are obese.*
- *Obesity is considered a global pandemic since more than 1.9 billion adults worldwide were considered overweight in 2014, and nearly 600 million of those adults were obese.*

FOOD AND DISEASE

Food is fuel for the body. When there is an excess of this fuel or too much of the wrong kind of fuel is consumed, health and well-being may decline. For example, obesity, hypertension, cardiovascular disease, diabetes, and certain types of cancers have been linked to dietary factors. Obesity is one of the biggest concerns because it is a precursor to many of the other diseases.

Obesity

Obesity is a medical condition in which the accumulation of excess adipose tissue (body fat) poses an adverse effect on health. According to the Centers for Disease Control and Prevention (CDC), approximately 35% of adults and 17% of children and adolescents are considered obese. Obesity increases the risk of developing hypertension, cardiovascular disease, type 2 diabetes, and certain types of cancers. One method used to measure a person's body fat composition is the body mass index.

Body Mass Index. The *body mass index (BMI)* is a calculation based on height and weight to determine approximate body fat composition. The CDC uses the BMI to define the following ranges:

- underweight — below 18.5
- normal weight — 18.5 to 24.9
- overweight — 25.0 to 29.9
- obese — 30.0 and above

Hypertension

Hypertension is a condition characterized by high blood pressure. Blood pressure is the pressure of blood within the arteries. Blood pressure varies throughout the day. However, hypertension results when blood pressure remains elevated. If uncontrolled, hypertension can lead to heart and kidney disease, as well as blindness. Factors such as high levels of sodium in the diet and being overweight have been shown to adversely affect blood pressure.

Cardiovascular Disease

Cardiovascular disease, also known as heart disease, is a class of medical conditions that affects the heart and blood vessels. Cardiovascular disease is the primary cause of death in the United States. Eating a diet rich in dietary fiber and low in saturated fats has shown to reduce the risk of certain types of cardiovascular disease. Maintaining an appropriate weight, taking part in physical activity, and abstaining from smoking are additional ways to diminish the threat of cardiovascular disease.

Diabetes

Diabetes, scientifically known as diabetes mellitus, is a group of diseases that impacts the metabolism due to insufficient insulin production (type 1) and/or high insulin resistance (type 2). Insulin is a hormone produced by the pancreas that is necessary for regulating blood sugar levels. Uncontrolled blood sugar levels can lead to serious ramifications such as amputations, blindness, heart and kidney disease, and premature death.

Cancers

Cancer is a group of diseases characterized by abnormal cell growth. Cancers can disrupt the normal functioning of organs and tissues due to the rapid multiplication of cancerous cells. Factors influencing the formation of cancer include obesity, physical inactivity, diets high in saturated fats, cigarette smoking, and a family history of cancer.

FEATURE 38, FRUITS AND VEGETABLES

Fruits and vegetables are naturally grown foods that are essential to human nutrition. Fruits provide a healthy source of carbohydrates and essential vitamins and minerals. Vegetables are also a healthy source of carbohydrates, and many contain fiber.

Unfortunately, most people do not get the recommend daily amount of fruits and vegetables. The Dietary Guidelines for Americans recommends at least four servings of fruit and five servings of vegetables per day. **See Figure 7-2.** Feature 38, Fruits and Vegetables, is a precondition for the New and Existing Interiors and New and Existing Buildings project types that requires that various types of fruits and vegetables be provided and promoted.

FRUITS AND VEGETABLES

FRUITS

VEGETABLES

Figure 7-2. Feature 38, Fruits and Vegetables, requires projects that sell food to offer a variety of fruit and vegetable choices and to promote fruit and vegetable consumption.

Part 1: Fruit and Vegetable Variety

Part 1: Fruit and Vegetable Variety of Feature 38 requires that a variety of fruit and vegetable choices be provided when food is sold or provided on a project's premises on a daily basis. When offered an assortment of fruits and vegetables, building occupants have a better opportunity to incorporate these healthier options in their diet. For Part 1, one of two requirements must be met. Part 1a requires a selection of at least two varieties of fruit and at least two varieties nonfried vegetables. Part 1b requires that at least 50% of all the available options be fruit (containing no added sugar) and/or nonfried vegetables.

Part 2: Fruit and Vegetable Promotion

Part 2: Fruit and Vegetable Promotion of Feature 38 concerns the promotion of healthier food options. Healthier foods can be promoted through the design of a cafeteria, visual displays or advertisements, and strategic placement of vegetables and fruits at the beginning of foodservice lines and at checkout locations. **See Figure 7-3.** For Part 2, if an owner-operated or contracted cafeteria is present, a project must be designed to include the following methods of promotion:

- visible and accessible salad bar or salad-providing section
- prominent visual displays of fruits and vegetables
- vegetables placed at the front of a foodservice line
- fruits placed at checkout locations

FRUIT AND VEGETABLE PROMOTION

Figure 7-3. A salad bar in an owner-operated or contracted cafeteria must be positioned in a visible and accessible location.

FEATURE 39, PROCESSED FOODS

A *processed food* is food that incurs a deliberate change before it is available for consumption. Examples of deliberate changes include cooking, freezing, dehydration, or milling. Although some processed foods, such as canned or frozen vegetables, have been minimally processed, many processed foods contain excessive levels of sugar, fat, sodium, refined flour, trans fat, or calories. Consumption of overly processed foods can contribute to health problems such as weight gain, diabetes, kidney disease, and high blood pressure. Feature 39, Processed Foods, is a precondition for all three project types that sets restrictions for processed foods.

BY THE NUMBERS

- *Even though the recommended limit for sugar intake is only 6 to 9 teaspoons per day, the average consumption in the United States is more than 22 teaspoons per day.*
- *For a quarter of the U.S. population, sugar-sweetened beverages (SSBs) such as fruit-flavored drinks, sodas, sports drinks, and energy drinks account for 200 calories consumed daily.*

Part 1: Refined Ingredient Restrictions

Part 1: Refined Ingredient Restrictions of Feature 39 primarily concerns sugar. High levels of added sugar are found in soft drinks, juices, candy, and desserts. **See Figure 7-4.** Three requirements, Part 1a–c, limit the availability of these goods or the amount of sugar in them. The fourth requirement, Part 1d, limits the access to goods with refined flour and promotes the use of whole-grain flour. To fulfill these four requirements, a project must meet the following limits:

- Beverages with more than 30 g of sugar per container may not be sold or provided by (or under contract with) the project owner, including the beverages in vending machines. Bulk containers of 1.9 L (2 qt) or larger are exempt from this requirement.
- At least 50% of the beverages must have 1 g of sugar or less per 16 mL (1.87 g or less per 1 oz).
- No non-beverage food item may contain more than 30 g of sugar per serving.
- In at least 50% of food offerings in which grain flour is the primary ingredient by weight, that ingredient must be a whole grain.

REFINED INGREDIENT RESTRICTIONS

Figure 7-4. Some processed foods contain excessive levels of sugar, fat, sodium, refined flour, trans fat, or calories.

Part 2: Trans Fat Ban

Trans fat, also known as a partially hydrogenated oil or trans-fatty acid, is an unsaturated fatty acid with hydrogen atoms on opposite sides of a double carbon bond, which makes its structure similar to a saturated fat. Trans fats are rare in natural foods but abundant in processed foods due to hydrogenation. Hydrogenation is the process of forcing hydrogen gas into oil at high pressure in order to increase the shelf life and prevent rancidity of the oil. Hydrogenation turns a liquid fat into a solid fat at room temperature.

Part 2: Trans Fat Ban of Feature 39 prohibits the sale or distribution of any foods or beverages that contain partially hydrogenated oil (trans fat). **See Figure 7-5.** Trans fat raises a body's unhealthy low-density lipoprotein (LDL) cholesterol and lowers healthy high-density lipoprotein (HDL) cholesterol.

TRANS FAT BAN

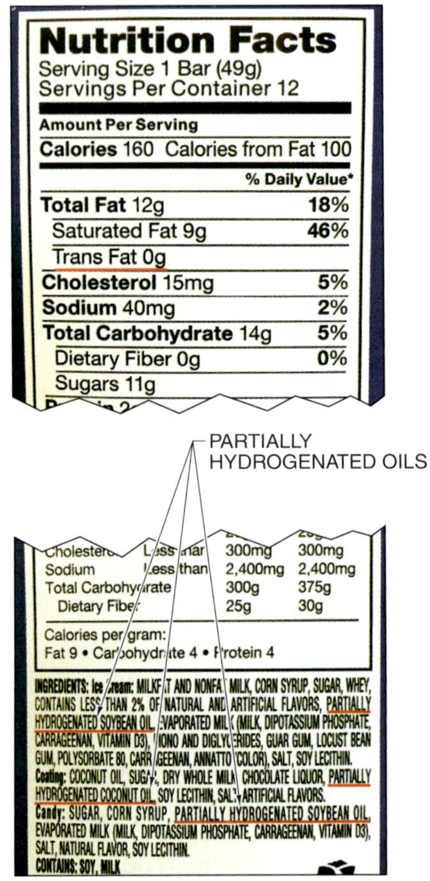

Figure 7-5. Although a label may list 0 g of trans fat, there still may be up to 0.5 g per serving if the ingredients list contains partially hydrogenated oil.

FEATURE 40, FOOD ALLERGIES

Consumption of or contact with food allergens can be very dangerous for people who have a food allergy or sensitivity. Allergic reactions, such as hives, vomiting, stomach cramps, shortness of breath, swelling of the tongue, dizziness, or shock, can occur immediately upon contact with food or several hours after exposure. Since 8% of children and 4% of adults in the United States have food allergies of some type, the U.S. Food and Drug Administration (FDA) requires that any food sold or distributed be properly labeled to identify any allergens that might be contained in the food, whether the food is prepacked or prepared on-site. **See Figure 7-6.** Feature 40, Food Allergies, is a precondition for all three project types and requires all foods to have allergen labels.

FOOD ALLERGEN LABELING

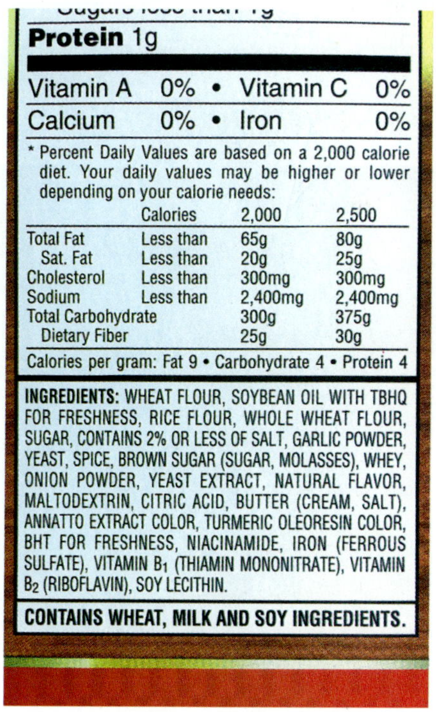

Figure 7-6. Food allergens must be properly addressed on any food that is sold or distributed on-site on a daily basis.

Part 1: Food Allergy Labeling

A person can be allergic to virtually any food or ingredient. However, there are eight food allergens that account for the majority of all food-based allergic reactions. In addition to these eight allergens, gluten sensitivity is a problem for a small percentage of the population. **See Figure 7-7.**

BY THE NUMBERS

- *Every year, approximately 30,000 people visit emergency rooms, 2000 are hospitalized, and 150 die due to allergic reactions to food.*

Food Allergies and Sources

Allergen	Potential Food Sources
Peanuts	• Chili sauce, hot sauce, mole sauce, marinades, glazes, and salad dressings • Pudding, cookies, crackers, egg rolls, and potato pancakes • Some vegetarian food products advertised as meat substitutes • Foods that contain extruded, cold-pressed, or expelled peanut oil
Fish	• Imitation fish or shellfish • Salad dressings, Worcestershire sauce, and sauces made with Worcestershire
Shellfish	• Many Asian dishes include fish sauce • Shellfish protein can become airborne in the steam released during cooking
Soy	• Baked goods, cereals, crackers, infant formula, soups, and sauces • Some brands of tuna and peanut butter • Soybean oil that has not been cold pressed, expeller pressed, or extruded
Milk and Dairy	• Many nondairy products contain casein (a milk derivative) • Some brands of tuna fish contain casein • Butter melted on top of grilled steaks (often not visible) or foods cooked in butter • Many baked products contain dry milk
Egg	• Egg whites are often used to create the foam topping on specialty coffee drinks • Egg wash is sometimes used on pretzels before they are dipped in salt • Some brands of egg substitutes contain egg whites • Most processed cooked pastas (including those in prepared soups) contain eggs or are processed on equipment shared with egg-containing pastas
Wheat	• Some brands of hot dogs, imitation crabmeat, and ice cream contain wheat • Some Asian dishes contain wheat flour shaped to look like beef, pork, or shrimp
Tree Nuts	• Salads and salad dressings, barbeque sauce, and honey • Meat-free burgers, fish dishes, pancakes, and pasta • Piecrust and the breading on meat, poultry, and fish
Gluten	• Flour-based products such as breads, cereals, pastas, cookies, and cakes • May be found in bouillon cubes, soy or teriyaki sauces, gravies, certain deli meats, and beer

Figure 7-7. Eight food allergens and gluten sensitivity account for the majority of food-based allergic reactions.

Gluten is a type of protein found in the endosperm of cereal grains such as wheat, rye, barley, and spelt. Gluten is prevalent in flour and flour-based products such as breads, cereals, pastas, and baked goods. It can also be found in other foods such as bouillon cubes, soy sauce, and deli meats.

Part 1: Food Allergy Labeling of Feature 40 requires labels for all foods sold or provided on a daily basis on the premises of a project. These labels are to be placed on packaging, menus, signage, or electronic media. The labels must clearly indicate whether the food contains the following allergens:

- peanuts
- fish
- shellfish
- soy
- milk and dairy products
- egg
- wheat
- tree nuts
- gluten

FEATURE 41, HAND WASHING

Feature 41, Hand Washing, is a precondition for the New and Existing Interiors and New and Existing Buildings project types. This feature requires that projects provide adequate hand washing facilities and supplies. Proper hand washing is an important method of limiting the transmission of pathogens (disease-causing agents) when handling food.

BY THE NUMBERS

- *In the United States alone, foodborne illnesses account for 48 million people getting sick with 128,000 hospitalizations and 3000 deaths.*

Part 1: Hand Washing Supplies

Part 1: Hand Washing Supplies of Feature 41 requires that adequate hand washing supplies—specifically fragrance-free soap and disposable paper towels—be provided at all sink locations. **See Figure 7-8.** The soap must meet the hand-hygiene product requirements of Table A4 in Appendix C of the WELL Building Standard. Disposable paper towels are more effective in removing bacteria than air dryers. Air dryers are not forbidden, but can only be used to supplement disposable paper towels.

HAND WASHING SUPPLIES

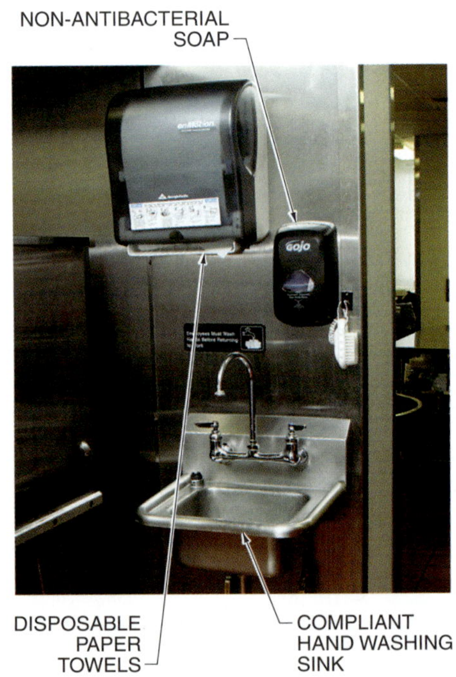

Figure 7-8. Proper hand washing supplies must be provided at all sink locations.

Part 2: Contamination Reduction

Part 2: Contamination Reduction of Feature 41 requires that either liquid soap dispensers with disposable and sealed soap cartridges or bar soap with a soap rack that allows for drainage be provided at every sink to reduce the possibility of contamination. Contamination risks occur with refillable bulk soap dispensers through airborne or physical germ transfer on the dispenser or contamination of the bulk soap itself. The nozzle of a refillable bulk soap dispenser is a permanent part of the soap-dispensing fixture and is at serious risk of pathogen transfer unless a strict cleaning plan is maintained.

Part 3: Sink Dimensions

Part 3: Sink Dimensions of Feature 41 sets minimum sink dimensions to ensure that there is enough clearance to wash hands without touching the sides of the basin or splashing water over the sides. Minimally, the column of water from the faucet to the bottom of the basin must be at least 25 cm (10″) in length, and the basin must be at least 23 cm (9″) in width and length. **See Figure 7-9.**

FEATURE 42, FOOD CONTAMINATION

Raw meats, fish, and poultry have an increased risk of being contaminated with harmful pathogens due to the original animals, the processing methods and equipment, or the environment. Some of these pathogens include E. coli, salmonella, listeria, and parasites, which can be killed by cooking foods to the recommended safe temperature. The proper storage and preparation of raw foods is also important for reducing the risk of pathogen contamination. Feature 42, Food Contamination, is a precondition for the New and Existing Interiors and New and Existing Buildings project types and requires proper cold storage spaces.

Part 1: Cold Storage

Part 1: Cold Storage of Feature 42 has two requirements. Part 1a specifies that if any raw meat, fish, or poultry is prepared or stored on-site, then separate storage must be available. At least one removable, cleanable drawer or container must be located at the bottom of the refrigerated unit and be properly labeled for storing raw meat, fish, and poultry. The container must be provided at the bottom of the unit to prevent cross contamination due to leaks. Part 1b requires a visual display of the refrigerated unit's holding temperature to ensure that the proper storage temperature is maintained.

FEATURE 43, ARTIFICIAL INGREDIENTS

There are hundreds of artificial ingredients that can be added to food products, including artificial colors, artificial flavors, artificial sweeteners, emulsifiers, preservatives, fillers, and other chemicals. Some of these artificial ingredients can pose a serious health risk to people, especially those with preexisting medical issues. Feature 43, Artificial Ingredients, is an optimization for the Core and Shell project type and a precondition

HAND WASHING SINK DIMENSIONS

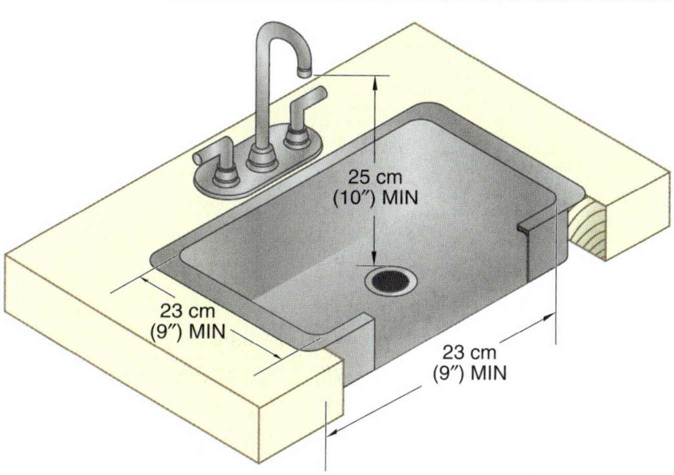

Figure 7-9. Part 3: Sink Dimensions of Feature 41, Hand Washing, requires that the column of water from a faucet to the bottom of a basin be at least 25 cm (10″) in length and that the basin be at least 23 cm (9″) in width and length.

for the New and Existing Interiors and New and Existing Buildings types that requires labeling for artificial ingredients.

Part 1: Artificial Substance Labeling

Although artificial ingredients may enhance the taste or physical properties of food, they provide no positive nutritional value and should be avoided. Part 1: Artificial Substance Labeling of Feature 43 does not prohibit artificial ingredients, but it does require that any food or beverage sold or provided on a daily basis by (or under contract with) the project owner be labeled to indicate if it has artificial ingredients. **See Figure 7-10.**

FEATURE 44, NUTRITIONAL INFORMATION

Without adequate nutritional information, consumers are unable to make informed decisions concerning their food choices. To protect consumers, the FDA requires nutritional labeling for most prepackaged foods. Feature 44, Nutritional Information, is an optimization for the Core and Shell project type and a precondition for the New and Existing Interiors and New and Existing Buildings types that requires detailed nutritional information to be provided for all packaged or prepared foods and beverages sold or provided on the premises.

Artificial Ingredient Uses and Health Concerns

Ingredients	Uses	Health Concerns
Artificial Colors	Added to soft drinks, breakfast cereals, candy, snack foods, baked goods, frozen desserts, and salad dressings as a colorant	Allergic reactions; cancers; hyperactivity; behavorial issues in children; headaches
Artificial Flavors	Added to soft drinks, candy, breakfast cereals, gelatin desserts, snack foods, and microwave popcorn as a flavoring	Allergic reactions; dermatitis; eczema; asthma
Artificial Sweeteners	Added to low- and zero-calorie diet foods and beverages as sweetener	Cancer; headaches; dizziness; impaired metabolic functions
Brominated Vegetable Oils	Added to soft drinks and energy drinks to keep citrus flavor from separating	Increased triglycerides and cholesterol; organ damage
Potassium Bromate	Used in baked goods to strengthen dough	Cancer
Butylated Hydroxyanisole (BHA)	Added to foods to protect fats from oxidizing	Suspected carcinogen in high doses
Butylated Hydroxytoluene (BHT)	Added to foods to protect fats from oxidizing	Suspected carcinogen in high doses
Monosodium Glutamate (MSG)	Added to foods as a flavor enhancer	Headaches; nausea; weakness; tingling or numbness in hands or face
Hydrolyzed Vegetable Protein (HVP)	Added to foods as a flavor enhancer (contains MSG)	Headaches; nausea; weakness; tingling or numbness in hands or face
Sodium Nitrate and Sodium Nitrite	Added to processed meats as a preservative, colorant, or flavoring	Heart disease; diabetes; cancer
Sulfites	Added to dried fruits, wine, and other foods as a preservative	Headache; rash; breathing issue for asthmatics

Figure 7-10. Feature 43, Artificial Ingredients, requires labeling for food or beverages that contain artificial ingredients since artificial ingredients provide no positive nutritional value and should be avoided.

Part 1: Detailed Nutritional Information

The detailed nutritional information that must be listed or displayed for all food and beverages that are sold or provided on a daily basis on the premises by (or under contract with) the project owner are similar to the FDA requirements for food labeling. Part 1: Detailed Nutritional Information of Feature 44 requires that the nutritional information of each meal, food, or beverage item be included on its packaging, menus, or other signage. **See Figure 7-11.** The following nutritional information must be accurately displayed:

- total calories
- macronutrient content, including total protein, fat, and carbohydrates (in weight and percent of estimated daily requirement)
- total sugar content

FEATURE 45, FOOD ADVERTISING

People are subjected to a variety of media influences that shape their food choices and eating behaviors. Both television and Internet advertisements consistently promote unhealthy food products aimed at consumers of all ages, especially children. Feature 45, Food Advertising, is an optimization for the Core and Shell project type and a precondition for the New and Existing Interiors and New and Existing Buildings project types that aims to counteract the negative nutritional messages in advertising and to promote healthier food options and eating habits.

WELLNESS FACT

The Nutrition Labeling and Education Act of 1990 is a U.S. federal law that requires nutrition information labels on all packaged food and beverages.

DETAILED NUTRITIONAL INFORMATION

Figure 7-11. Detailed nutritional information, such as calories, macronutrient and micronutrient content, and sugar content must be listed or displayed for all food and beverages that are sold or distributed on-site.

Part 1: Advertising and Environmental Cues

Part 1: Advertising and Environmental Cues of Feature 45 restricts the promotion of unhealthy food products. Part 1 states that a food or beverage item may not be advertised on a project's premises if the item fails to meet the requirements set forth in Feature 39, Processed Foods. For example, if a food item contains trans fat, it may not be advertised anywhere on the premises.

Part 2: Nutritional Messaging

Part 2: Nutritional Messaging of Feature 45 encourages healthy eating habits through promotional displays such as educational posters, brochures, and other visual media. Specifically, Part 2 requires the prominent display of at least three instances of messaging per project that achieve either or both of the following:

- encourage the consumption of whole, natural foods
- discourage the consumption of sugary or processed foods and beverages

FEATURE 46, SAFE FOOD PREPARATION MATERIALS

Most food preparation materials sold today are generally considered safe when properly used and maintained. However, certain materials, such as plastics, nonstick cookware, and some metals can be hazardous if used improperly. Some plastics and nonstick cookware can release toxic chemicals when heated too high. When heated or exposed to acidic ingredients, utensils and cookware made from uncoated metals such as aluminum or copper can allow excessive amounts of metal to leach into food. There is also a risk of porous surfaces harboring harmful pathogens that can be passed on to the food being prepared. Feature 46, Safe Food Preparation Materials, is an optimization for the New and Existing Interiors and New and Existing Buildings project types.

Part 1: Cooking Material

Part 1: Cooking Material of Feature 46 requires that any cooking tools with the exception of cutting boards be made entirely of one or more of the following materials:

- lead-free ceramics
- cast iron
- stainless steel
- glass
- coated aluminum
- solid wood, either untreated or treated with food-grade mineral or linseed oil

Part 2: Cutting Surfaces

Cutting surfaces are at risk of cross-contamination from raw ingredients such as meat, poultry, or fish. If the cutting surface is made from a porous material, or if the surface is excessively worn or grooved, bacteria can find a place to hide and multiply. Part 2: Cutting Surfaces of Feature 46 requires that all cutting boards be made of one of the following materials:

- marble
- plastic
- glass
- pyroceramics
- solid wood, either untreated or treated with food-grade mineral or linseed oil

FEATURE 47, SERVING SIZES

People in developed countries, especially the United States, have seen the portion sizes of food increase steadily over the years. As portion sizes have gotten bigger, obesity rates and other weight-related diseases have increased. People have been trained over the years to "clean their plate." However, this behavior can lead to an excessive calorie intake due to larger portion sizes. Feature 47, Serving Sizes, is an optimization for the New and Existing Interiors and New and Existing Buildings project types that calls for smaller portion sizes and sets size limitations for dishware.

BY THE NUMBERS

- *The average person in the United States consumed nearly 2600 calories per day in 2010 versus about 2100 calories per day in 1970.*

Part 1: Meal Sizes

Part 1: Meal Sizes of Feature 47 applies to projects where sold or provided food is prepared to order. Part 1 requires that lower calorie portions or versions be offered for at least half of the main course options on a menu. These portions must be less than 650 kcal (650 Cal) and cost less than the regular or full-size version. They can be smaller serving sizes or have alternative ingredients. The lower-calorie versions must be clearly marked on the menu.

Part 2: Dishware

Although people have become accustomed to seeing plates overfilled with food and often perceive larger portions as a better value, presenting food on smaller plates is one way to alter this perception and encourage portion control. Limiting the size of the dishware gives the impression that the plate is still full and that there is perceived value. Part 2: Dishware of Feature 47 sets size limitations for dishware on the premises of projects that sell or distribute self-served food on a daily basis. **See Figure 7-12.** Size limitations for self-serve dishware include the following:

- circular plates with diameters no larger than 25 cm (10″)
- noncircular plates with surface areas no larger than 507 cm^2 (79 in^2)
- bowls no larger than 473 mL (16 oz)
- cups no larger than 473 mL (16 oz)

FEATURE 48, SPECIAL DIETS

Many people have dietary restrictions. These restrictions may be due to health-related issues, such as a peanut allergy, or personal or religious beliefs. Feature 48, Special Diets, is an optimization for the New and Existing Interiors and New and Existing Buildings project types that requires alternative food options be provided to address special dietary needs.

Part 1: Food Alternatives

Part 1: Food Alternatives of Feature 48 requires that meals or catering provided by (or under contract with) a project owner have at least one main course option for each of the following food allergies or dietary restrictions as necessary or by request:

- peanut-free
- gluten-free
- lactose-free
- egg-free
- no animal, seafood, or dairy products
- no animal products except eggs or dairy

FEATURE 49, RESPONSIBLE FOOD PRODUCTION

Interest in sustainable practices has led many people to incorporate organic and humanely raised foods into their diets. *Organic food* is food produced without the use of chemically formulated fertilizers, growth stimulants, antibiotics, pesticides, or spoilage-inhibiting radiation. A *humanely raised food* is a meat, egg, or dairy product from an animal that has been kept, fed, and processed according to voluntary humane animal welfare standards. Feature 49, Responsible Food Production, is an optimization for the New and Existing Interiors and New and Existing Buildings project types that requires the adoption of organic and humane foods.

Part 1: Sustainable Agriculture

Part 1: Sustainable Agriculture of Feature 49 requires that all produce sold or provided on the premises on a daily basis by (or under contract with) the project owner have a federally certified organic label based on country. In the United States, the U.S. Department of Agriculture (USDA) sets standards for organic food products through the Organic Foods Production Act (OFPA) and the National Organic Program (NOP). **See Figure 7-13.**

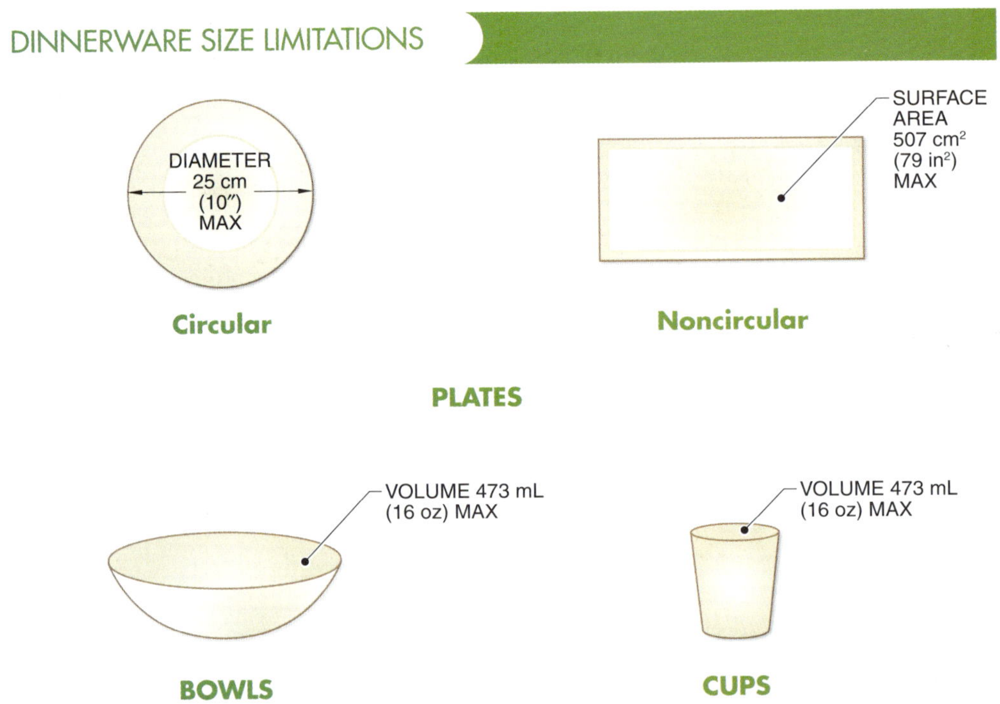

Figure 7-12. For Part 2: Dinnerware Sizes of Feature 47, Serving Sizes, any dinnerware must meet the listed size requirements.

Figure 7-13. Organic produce that meets USDA standards for organic food products can bear the "USDA ORGANIC" label. Humane agriculture accommodates the health and natural behavior of livestock.

Part 2: Humane Agriculture

Part 2: Humane Agriculture of Feature 49 requires that all meat, egg, and dairy products sold or provided on the premises of a project on a daily basis by (or under contract with) the project owner meet criteria for the humane treatment of livestock. This means that the products must have both a Certified Humane™ label and a federally certified organic label, or equivalent.

FEATURE 50, FOOD STORAGE

Maintaining proper temperature and humidity is important for keeping fruits and vegetables (produce) at their peak freshness and to reduce spoilage. When these food items are kept in optimal conditions, they maintain their quality and nutritional value. Feature 50, Food Storage, is an optimization for the New and Existing Interiors and New and Existing Buildings project types that requires a project to have sufficient cold storage space for produce in the form of refrigerators and/or other food storage equipment for use by the building occupants.

Part 1: Storage Capacity

Part 1: Storage Capacity of Feature 50 requires either a minimum of 20 L (0.7 ft^3) of storage space per regular building occupant, with a maximum of 7000 L (247 ft^3), or evidence that the storage volume exceeds occupant demand by at least 20%. For example, a building with 125 regular building occupants would require a minimum of 2500 L (87.5 ft^3) of cold storage space. The storage space, whether a refrigerator or other piece of food storage equipment, must include temperature control capabilities.

FEATURE 51, FOOD PRODUCTION

Providing a nearby or on-site garden or greenhouse is an effective way to encourage occupants to select healthy food options. An on-site garden or greenhouse can be located outside of a building, on its roof, or inside the building. Feature 51, Food Production, is an optimization for all three project types that requires a sufficiently sized gardening space as well as the equipment and supplies to produce edible fruits and vegetables. **See Figure 7-14.**

NEARBY OR ON-SITE FOOD PRODUCTION

National Garden Bureau Inc.

Figure 7-14. Providing a nearby or on-site garden or greenhouse is an effective way to encourage occupants to select healthy food options.

Part 1: Gardening Space

Part 1: Gardening Space of Feature 51 requires a gardening space of sufficient size to provide ample produce for building occupants. This gardening space should consist of a garden or greenhouse with food-bearing plants, or edible landscaping such as fruit trees or herbs. There must be a minimum of at least 0.1 m^2 (1 ft^2) per regular occupant, with a maximum of 70 m^2 (754 ft^2). The growing space must be within a distance of 0.8 km (0.5 miles) from the project boundary.

Part 2: Planting Support

Part 2: Planting Support of Feature 51 requires that any gardening spaces have adequate supplies. Gardening supplies include the following:

- planting medium (such as soil, soil-less mixes, or mulch)
- irrigation
- lighting for indoor spaces
- plants
- gardening tools

Messermeister

Gardening not only provides healthy food options but can also help alleviate stress and improve mood.

FEATURE 52, MINDFUL EATING

Distracted eating, such as eating alone at a workstation, has become more common due to busy workdays and uninviting break rooms. Distracted eating is a poor eating habit because it may lead to increased calorie intake. Feature 52, Mindful Eating, is an optimization for all three project types that aims to supply occupants with eating areas and break rooms that are conducive to the attentive consumption of food. These spaces include the area itself as well as the furnishings contained within the space.

Part 1: Eating Spaces

Part 1: Eating Spaces of Feature 52 defines the size and location of eating spaces within a building. Eating spaces must have enough seating to accommodate at least 25% of total occupants at a given time. **See Figure 7-15.** Occupants who eat in these areas may have improved eating habits, social interactions, and stress levels.

EATING SPACES

Figure 7-15. Feature 52, Mindful Eating, requires eating spaces to have enough seating to accommodate at least 25% of total occupants at a given time.

Part 2: Break Area Furnishings

Part 2: Break Area Furnishings of Feature 52 specifies which furnishings must be included in all occupant eating spaces. All eating spaces must include the following furnishings:

- a refrigerator, a device for reheating of food (such as a microwave or toaster oven), and a sink
- amenities for dishwashing
- at least one cabinet or storage unit available for employee use
- eating utensils, including spoons, forks, knives, and microwave-safe plates and cups

KEY TERMS AND DEFINITIONS

body mass index (BMI): A calculation based on height and weight to determine approximate body fat composition.

cancer: A group of diseases characterized by abnormal cell growth.

carbohydrate: Any of a group of organic compounds that includes sugars, starches, celluloses and gums and serves as a major energy source to support bodily functions and physical activity.

cardiovascular disease: A class of medical conditions that affects the heart and blood vessels. Also known as heart disease.

diabetes: A group of diseases that impacts the metabolism due to insufficient insulin production (Type 1) and/or high insulin resistance (Type 2). Scientifically known as diabetes mellitus.

gluten: A type of protein found in the endosperm of cereal grains such as wheat, rye, barley, and spelt.

humanely raised food: A meat, egg, or dairy product from an animal that has been kept, fed, and processed according to voluntary humane animal welfare standards.

hypertension: A condition characterized by high blood pressure.

lipid: An energy-providing nutrient made from fatty acids.

macronutrient: A nutrient needed by the body in large amounts.

malnutrition: A condition that results from insufficient nutrient intake, excess nutrient intake, or nutrient intake in the wrong proportions.

mineral: An inorganic nutrient that is required in small amounts to help regulate body processes.

nutrient: A chemical that is required for metabolic processes, which must be taken from food or another external source.

obesity: A medical condition in which the accumulation of excess adipose tissue (body fat) poses an adverse effect on health.

organic food: Food produced without the use of chemically formulated fertilizers, growth stimulants, antibiotics, pesticides, or spoilage-inhibiting radiation.

processed food: Food that incurs a deliberate change before it is available for consumption.

protein: An energy-providing nutrient that is made of carbon, hydrogen, oxygen, and nitrogen assembled in chains of amino acids.

trans fat: An unsaturated fatty acid with hydrogen atoms on opposite sides of a double carbon bond, which makes its structure similar to a saturated fat. Also known as partially hydrogenated oil or trans-fatty acid.

vitamin: An organic nutrient that is required in small amounts to help regulate body processes.

CHAPTER REVIEW

Completion

_______________ 1. A(n) ___ is a nutrient needed by the body in large amounts.

_______________ 2. Proteins and carbohydrates are examples of ___.

_______________ 3. ___ is a medical condition in which the accumulation of excess adipose tissue (body fat) poses an adverse effect on health.

_______________ 4. A body mass index (BMI) of ___ or more may indicate an individual is overweight or obese.

_______________ 5. Part 1: Refined Ingredient Restrictions of Feature 39, Processed Foods, primarily concerns ___.

_______________ 6. Proper hand washing is an important method of limiting the transmission of ___ when handling food.

_______________ 7. Although ___ may enhance the taste or physical properties of food, they provide no positive nutritional value and should be avoided.

_______________ 8. Feature 45, ___, aims to counteract the negative nutritional messages delivered by advertising and promote healthier food options and eating habits.

_______________ 9. The reduced portion size of sold or provided food under Part 1: Meal Sizes of Feature 47, Serving Sizes, must be less than ___ kcal and cost less than the regular or full-size version.

_______________ 10. ___ food is food produced without the use of chemically formulated fertilizers, growth stimulants, antibiotics, pesticides, or spoilage-inhibiting radiation.

_______________ 11. The requirement of Part 1: Storage Capacity of Feature 50, Food Storage, can be met by providing a minimum of ___ of storage space per regular building occupant.

_______________ 12. The growing space for Part 1: Gardening Space of Feature 51, Food Production, must be within a distance of ___ from the project boundary.

_______________ 13. For Part 1: Eating Spaces of Feature 52, Mindful Eating, eating spaces must have enough seating to accommodate at least ___% of total occupants at a given time.

Short Answer

1. List the six essential nutrients and explain the importance of nutrients in terms of the aims of the Nourishment Concept.

2. What five major negative health conditions are related to dietary factors?

CHAPTER REVIEW *(continued)*

3. List the four design interventions that are covered under Part 2: Fruit and Vegetable Promotion of Feature 38, Fruits and Vegetables.

4. List the nine ingredients that are required to be labeled by Part 1: Food Allergy Labeling of Feature 40, Food Allergies.

5. What are the minimum sink dimensions that are required for Part 3: Sink Dimensions of Feature 41, Hand Washing?

6. List the information that must be displayed on packaging, menus, or other signage for Part 1: Detailed Nutritional Information of Feature 44, Nutritional Information.

7. List the names of the two parts of Feature 46, Safe Food Preparation Materials.

8. List the maximum dishware sizes that are set by Part 2: Dishware of Feature 47, Serving Sizes.

9. List the food allergies or dietary restrictions for which a main course option must be provided for meals or catering under Part 1: Food Alternatives of Feature 48, Special Diets.

10. List the gardening supplies required by Part 2: Planting Support of Feature 51, Food Production.

WELL AP EXAM PRACTICE QUESTIONS

1. What does Feature 43, Artificial Ingredients, require of any food or beverage that is sold or provided on-site on a daily basis?

A. Must not contain gluten
B. Must not contain any artificial ingredients
C. Be labeled to indicate total calories and trans fat
D. Be labeled to indicate if it has artificial ingredients

2. Part 1: Detailed Nutritional Information of Feature 44, Nutritional Information, requires that nutrition labels contain information on which nutrients?

A. Calcium, magnesium, iron, vitamins A and C, and total sugar content
B. Total calories, allergens, and total sugar content
C. Total calories, macronutrients, and total sugar
D. Total calories; vitamins A, C, and K; fats; and carbohydrates

3. Which Nourishment Concept feature requires an architectural drawing for an annotated document?

A. Part 1: Meal Sizes of Feature 47, Serving Sizes
B. Part 1: Sustainable Agriculture of Feature 49, Responsible Food Production
C. Part 2: Planting Support of Feature 51, Food Production
D. Part 1: Eating Spaces of Feature 52, Mindful Eating

4. What is the best strategy to encourage healthy eating choices?

A. Use educational posters and advertising to encourage the consumption of whole, natural foods and cuisines.
B. Use educational posters and advertising to encourage the consumption of sugary or processed foods, beverages, and snacks.
C. Place vegetable dishes at the end of a foodservice line.
D. Serve meals on circular plates larger than 25 cm (10″).

5. What are the best strategies to reduce the chances of foodborne illnesses?

A. Proper hand washing for food preparers and storing raw meat, fish, and poultry in a separate drawer or container at the top of a cold storage unit
B. Properly cooking foods to their recommended temperature and cleaning the floors and counters with antibacterial cleaners
C. Keeping all raw meat, fish, and poultry at 3° below room temperature
D. Proper hand washing for food preparers and storing raw meat, fish, and poultry in a separate drawer or container at the bottom of a cold storage unit

CHAPTER EIGHT

LIGHT

All work and living spaces require light for human health and activity. Artificial light sources are often used to provide illumination in the built environment, sometimes to the exclusion of natural light from the sun. Sunlight should be a significant source of illumination in the built environment. The type and quality of light can greatly affect visual acuity and human health, as well as influence circadian rhythms. The goal of the Light Concept in the WELL Building Standard is to optimize light sources in the built environment and avoid disruptions to circadian rhythms.

OBJECTIVES

- Identify the features in the Light Concept as preconditions or optimizations.
- Identify the different light sources, types of lighting, and lighting qualities.
- Describe the positive and negative effects of light on the human body.
- Explain the roles of the different photoreceptive cells in picking up visual lux or melanopic lux.
- Explain the role of light in managing circadian rhythms.
- Identify the causes of glare.
- Describe strategies for controlling daylight and artificial light in a space.
- Explain how daylight modeling and light calculations are used to control the quality and type of light entering a space.
- Describe the methods and tools used for validating lighting conditions.

KEY TERMS

- accent lighting
- ambient lighting
- annual sunlight exposure (ASE)
- arc tube
- circadian rhythm
- color rendering index (CRI)
- correlated color temperature (CCT)
- daylight modeling
- daylight window
- electromagnetic radiation
- electromagnetic spectrum
- equivalent melanopic lux (EML)
- fluorescent lamp
- glare
- high-intensity discharge (HID) lamp
- illuminance
- incandescent lamp
- lamp
- light-emitting diode (LED) lamp
- light reflectance value (LRV)
- luminaire
- luminance
- luminous flux
- luminous intensity
- spatial daylight autonomy (sDA)
- task lighting
- variable opacity glazing
- view window
- visible transmittance (VT)
- visual acuity
- wavelength (λ)

Learner Resources

ATPeResources.com/QuickLinks
Access Code: 935534

LIGHT CONCEPT

Light allows people to see, work, and interact with the environment around them. Too little light can make it difficult to perform daily tasks, while too much light or the wrong kind of light can cause distracting glare or disrupt the body's circadian rhythm. WELL aims to minimize disruptions to the body's circadian system, enhance productivity, support good sleep quality, and provide appropriate visual acuity where needed.

The Light Concept in WELL includes 11 features that provide illumination guidelines for avoiding disruptions to the circadian rhythm and productivity losses due to poor lighting quality and design. Core and Shell projects must meet one precondition but four optimizations can be pursued for higher certification. New and Existing Interiors and New and Existing Buildings must meet four preconditions, but seven optimizations can be pursued for higher certification. **See Figure 8-1.**

WELLNESS FACT

The Illuminating Engineering Society (IES) is a technical organization that publishes nationally and internationally used standards, such as The Lighting Handbook, for the lighting industry. These standards often take human health into consideration.

Natural light is ideal for workspace illumination and helps create a positive work environment.

LIGHT BASICS

In order to understand the requirements of many features in the Light Concept, it is necessary to understand the basics of light and lighting attributes. Light is a small but very important part of the electromagnetic spectrum. The *electromagnetic spectrum* is the range of all types of electromagnetic radiation. *Electromagnetic radiation* is energy in the form of electromagnetic waves, including gamma rays, X-rays, ultraviolet (UV) light, visible light, infrared (IR) light, microwave radiation, and radio waves.

Electromagnetic radiation is generally classified by wavelengths, using nanometers (nm). *Wavelength* (λ) is the distance between two points on a wave in which the wave repeats itself. The wavelengths of visible light range from roughly 400 nm to 700 nm. **See Figure 8-2.**

The human eye is not capable of seeing electromagnetic radiation with wavelengths outside the visible spectrum. However, the effects of other wavelengths of electromagnetic radiation, such as UV light and IR light, can still be detected. UV light can cause skin and eye damage and may fade carpet or furniture. IR light can cause air or objects to heat up, raising the temperature of a building.

Natural light is a great source of illumination that has a positive effect on human health. However, too much direct sunlight can cause excessive glare and affect visual acuity. Conversely, walls and rooftops in the built environment block much of the natural light. Therefore, a balance must be reached between too much natural light and too little, which may require alternative lighting sources.

Lighting Sources

Several different types of artificial lighting sources, or lamps, can be used to illuminate the built environment through luminaires. A *lamp* is an output component that converts electrical energy into light. A *luminaire* is a complete lighting unit that includes the components that distribute light, position and protect the lamps, and provide connection to the power supply.

The different types of lamps have different attributes, advantages, and disadvantages. The most common types of lamps used in the built environment are incandescent lamps, fluorescent lamps, high-intensity discharge (HID) lamps, and light-emitting diode (LED) lamps. **See Figure 8-3.**

WELL Building Standard Features: Light Concept						
Features	Project Type			Verification Documentation		
	Core and Shell	New and Existing Interiors	New and Existing Buildings	Letter of Assurance	Annotated Documents	On-Site Assessment
53, Visual Lighting Design						
Part 1: Visual Acuity for Focus	–	P	P	Architect	Policy Document	Spot Measurement
Part 2: Brightness Management Strategies	–	P	P		Professional Narrative	
54, Circadian Lighting Design						
Part 1: Melanopic Light Intensity for Work Areas	–	P	P	Architect		Spot Measurement
55, Electric Light Glare Control						
Part 1: Luminaire Shielding	–	P	P	Architect		
Part 2: Glare Minimization	P	P	P	Architect		
56, Solar Glare Control						
Part 1: View Window Shading	O	P	P	Architect		Spot Check
Part 2: Daylight Management	O	P	P	Architect		Spot Check
57, Low-Glare Workstation Design						
Part 1: Glare Avoidance	–	O	O			Visual Inspection
58, Color Quality						
Part 1: Color Rendering Index	–	O	O	Architect		
59, Surface Design						
Part 1: Working and Learning Area Surface Reflectivity	–	O	O	Architect		
60, Automated Shading and Dimming Controls						
Part 1: Automated Sunlight Control	–	O	O	MEP		
Part 2: Responsive Light Control	–	O	O	MEP		
61, Right to Light						
Part 1: Lease Depth	O	O	O		Architectural Drawing	Spot Check
Part 2: Window Access	–	O	O		Architectural Drawing	Spot Check
62, Daylight Modeling						
Part 1: Healthy Sunlight Exposure	O	O	O		Modeling Report	
63, Daylighting Fenestration						
Part 1: Window Sizes for Working and Learning Spaces	O	O	O		Architectural Drawing	Spot Check
Part 2: Window Transmittance in Working and Learning Areas	O	O	O	Architect		
Part 3: Uniform Color Transmittance	O	O	O	Architect		

Figure 8-1. The Light Concept in WELL includes 11 features that promote strategies such as proper lighting design and light levels to minimize disruptions to the circadian system, enhance productivity, and provide appropriate visual acuity.

Incandescent Lamps. An *incandescent lamp* is a lamp that produces light by the flow of current through a tungsten filament inside a sealed glass bulb sometimes filled with a gas. Incandescent lamps have a low initial cost and are simple to install and service. However, incandescent lamps have a lower electrical efficiency and shorter life than other lamps.

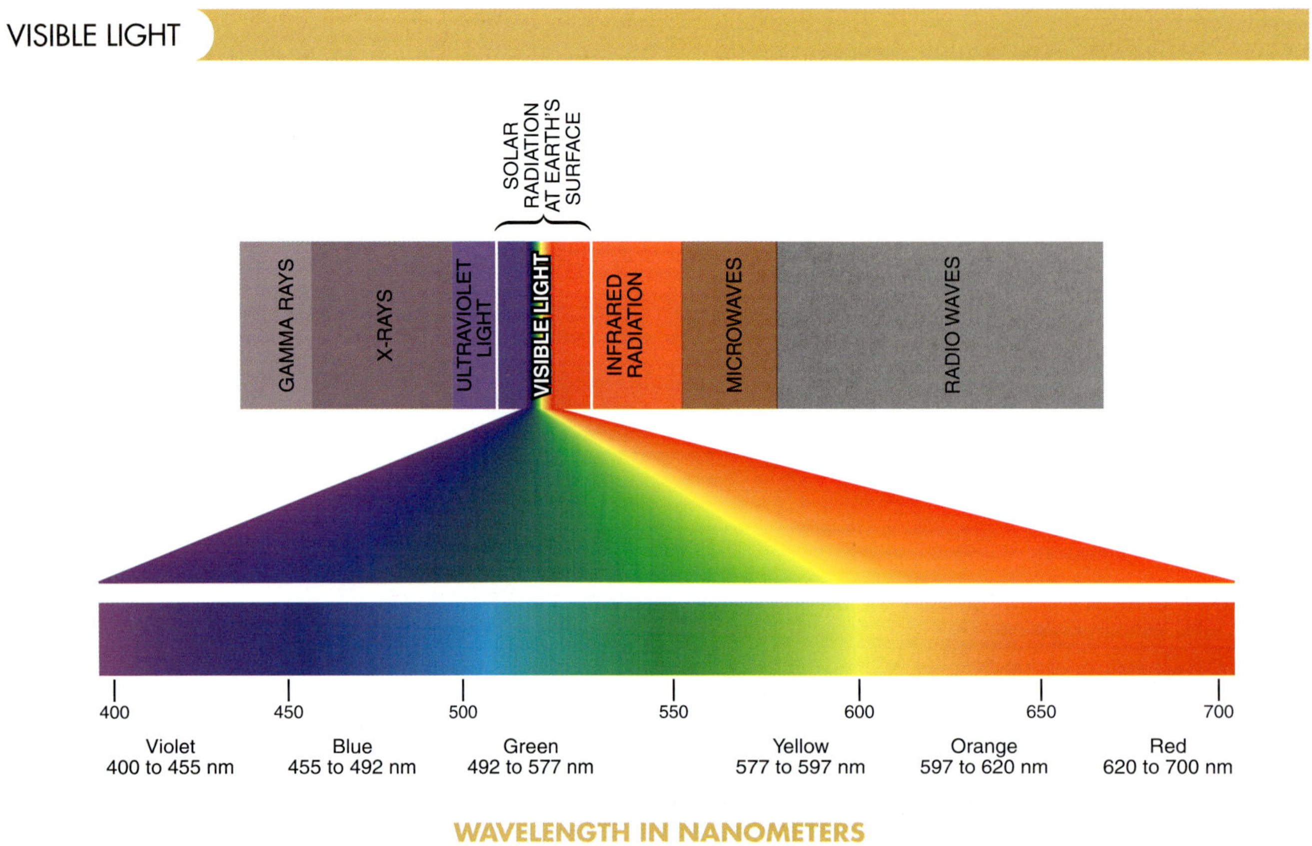

Figure 8-2. The wavelengths of visible light range from 400 nm to 700 nm.

Figure 8-3. The four most common lamp types are incandescent lamps, fluorescent lamps, high-intensity discharge (HID) lamps, and light-emitting diode (LED) lamps.

Fluorescent Lamps. A *fluorescent lamp* is a low-pressure discharge lamp in which ionization of mercury vapor transforms UV light generated by the discharge into visible light. A fluorescent lamp contains a mixture of inert gas (normally argon and mercury vapor), which is bombarded by electrons from its cathode. This provides UV light. The UV light causes the fluorescent material on the inner surface of the bulb to emit visible light.

High-Intensity Discharge Lamps. A *high-intensity discharge (HID) lamp* is a lamp that produces light from an arc tube. An *arc tube* is the light-producing element of an HID lamp. The arc tube contains metallic and gaseous vapors and electrodes. An arc is produced in the tube between the electrodes. The arc tube is enclosed in a bulb, which may include a phosphor or a diffusing coating that improves color rendering, increases light output, and reduces surface brightness. Types of HID lamps include low-pressure sodium, mercury-vapor, metal-halide, and high-pressure sodium lamps.

Light-Emitting Diode Lamps. A *light-emitting diode (LED) lamp* is a lamp with semiconductor devices that produce visible light when an electrical current passes through them. LED lamps use less energy and generate significantly less heat than other light sources. There is no warm-up period needed for an LED lamp. There are many LED lamp designs. For example, some fixtures contain modules that have a combination of red, green, and blue LEDs. This combination of colors allows a fixture to be tunable, which means that the color of light can be changed via an external control.

Types of Lighting

Lighting in the built environment can be classified as ambient lighting, task lighting, or accent lighting. **See Figure 8-4.** Effective lighting design includes the proper use of these three lighting classes while meeting the requirements of the building occupants and maintaining energy efficiency.

LIGHTING TYPES

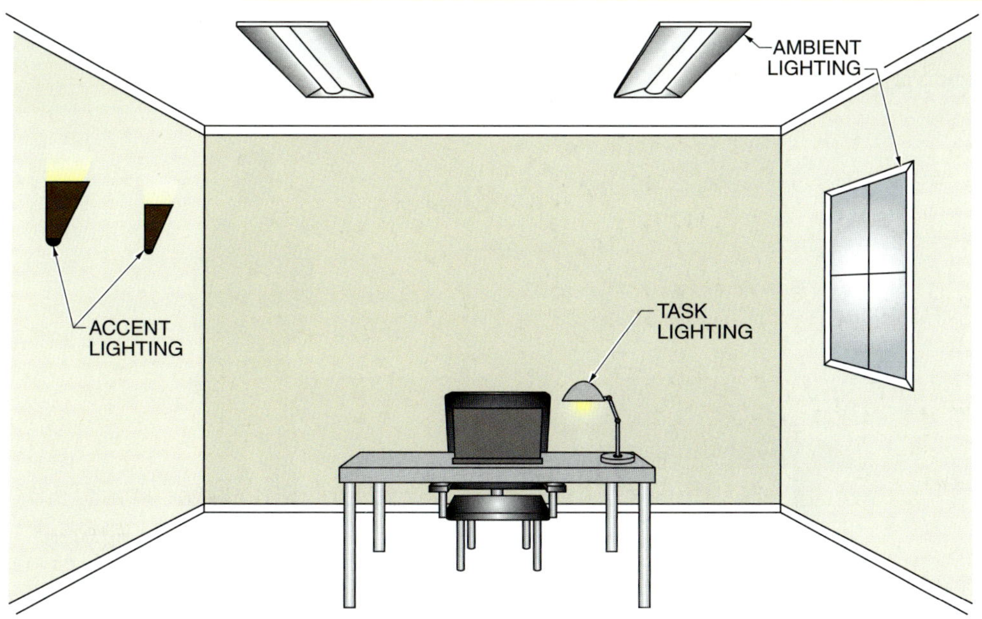

Figure 8-4. Lighting in the built environment can be classified into one of three types: ambient lighting, task lighting, or accent lighting.

Ambient Lighting. *Ambient lighting,* also known as general lighting, is the main source of nonspecific illumination in a space. Ambient lighting should be bright enough to allow safe movement around the space but not too bright as to cause visual discomfort. Ambient lighting should be integrated with a daylight control system to maintain the proper light levels in the spaces.

Task Lighting. *Task lighting* is lighting that is used to illuminate an area in order to allow the performance of a specific function. Task lighting is often controlled by building occupants. This level of control may be mandated by applicable building codes. Task lighting can reduce the amount of glare on reflective surfaces, thereby reducing eye strain and increasing productivity. The use of task lighting can also reduce the need for ambient lighting, which reduces energy costs.

Accent Lighting. *Accent lighting,* also known as decorative lighting, is lighting that is used to highlight architectural or design features of a space or to add visual interest to a space. Accent lighting is generally a small percentage of the lighting load but may be used to influence a positive effect on visual acuity and occupant health.

BY THE NUMBERS

- *Typically, an ambient light level of 300 lux is sufficient for most tasks.*

Lighting Qualities and Metrics

Lighting can be described based on the output of visible light by a luminaire or lighting system. The lighting quality and metrics of visible light can differ between different types of luminaires and lighting systems. Lighting qualities include color temperature and the absence of flicker. Lighting metrics include luminous flux, luminous intensity, illuminance, and luminance. **See Figure 8-5.**

Color Temperature. The color temperature of light is the relative color of light produced by a light source. This is not colored light, such as red, yellow, or green, but rather the perceived shade of white light. The color temperature of light is known as the correlated color temperature.

LIGHTING METRICS

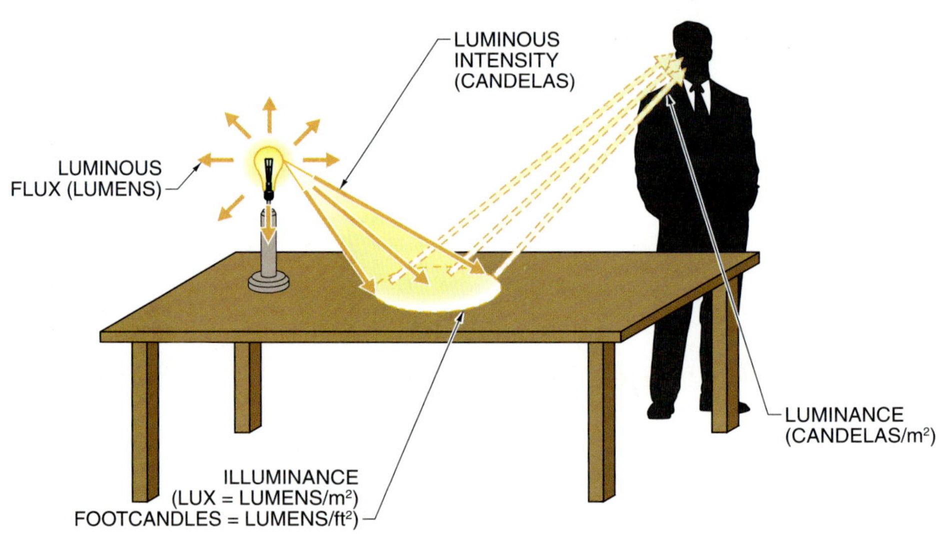

Figure 8-5. Lighting metrics include luminous flux, luminous intensity, luminance, and illuminance.

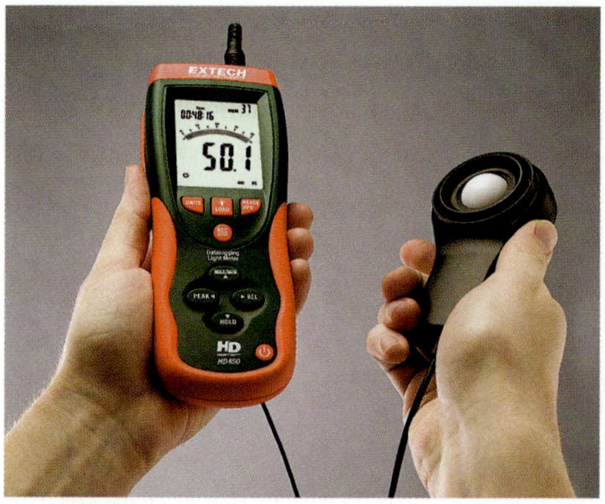

Extech Instruments

A handheld color temperature meter is used to measure the correlated color temperature of a light source.

Correlated color temperature (CCT) is the spectral distribution of electromagnetic radiation of a blackbody at a given temperature, measured in degrees Kelvin (K). The lower the value in degrees Kelvin, the "warmer" the light is perceived. The higher the value in degrees Kelvin, the "cooler" the light is perceived. Warmer light has more of a yellow shade, while cooler light has a blue shade. **See Figure 8-6.** Lighting manufacturers often supply the color temperature of their lamps or LEDs. If the temperature is not known, it can be read with a handheld color temperature meter.

Light Flicker. Flicker is the modulation, or change of intensity, of light output. Flicker is caused by the fluctuation of voltage. This occurs most frequently with fluorescent lights with older ballasts. Some older incandescent lights or LED lights with incorrect drivers or incompatibility with older dimmers may experience flicker as well. Flicker is an undesirable lighting quality that can cause headaches, eyestrain, or eye discomfort. Flicker can also be a distraction that may reduce worker productivity or cause an unsafe condition.

Luminous Flux. *Luminous flux* is the total luminous output of a light source, measured in lumens and weighted to the visual sensitivity of the human eye. Luminous flux is the total perceived amount of light being given off in all directions.

Luminous Intensity. *Luminous intensity* is radiant power weighted to human vision that describes light emitted by a source in a particular direction. Luminous intensity is measured by the candela.

Illuminance. *Illuminance* is the amount of light incident on a given surface measured in lux or footcandles. One lux is equivalent to one lumen per square meter. A footcandle (fc) is equivalent to one lumen per square foot. One footcandle is approximately 10.8 lux.

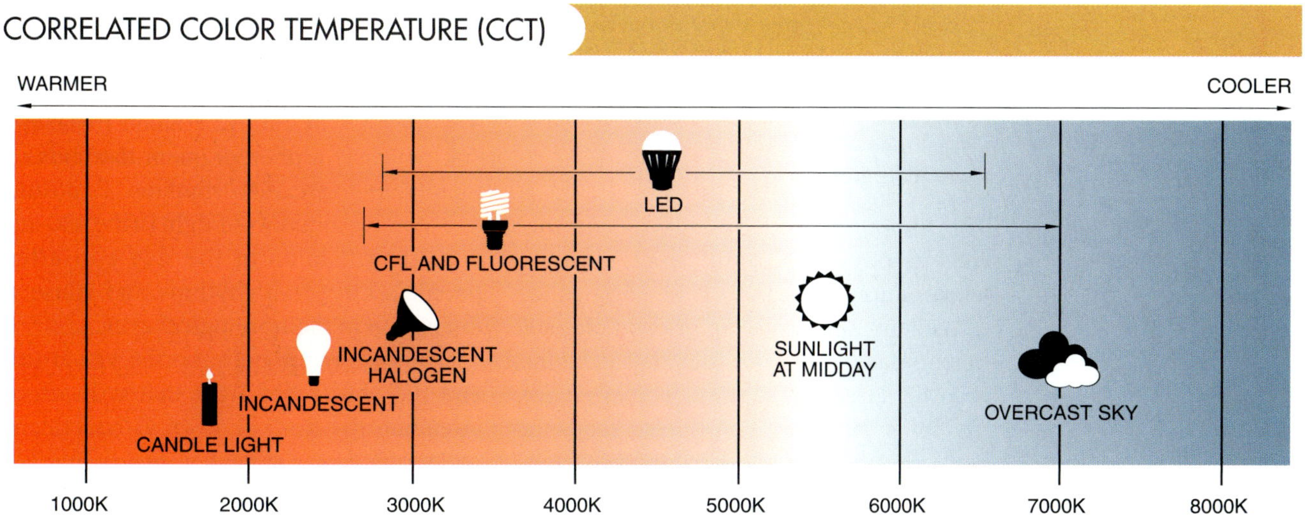

Figure 8-6. CCT is the relative color of light, warm or cool, produced by a light source.

Luminance. *Luminance* is a measurement of how bright a surface or light source will appear to the eye, measured in candela per square meter (cd/m^2) or footlamberts. Balanced luminance is important to maintain to avoid glare and visual discomfort.

LIGHT AND THE HUMAN BODY

A well-designed lighting system that combines sufficient levels of natural and artificial lighting can have a positive impact on the health and productivity of building occupants. Quality lighting can help alleviate fatigue, increase concentration, and increase productivity. Good lighting also makes for a safer environment as well. However, insufficient lighting levels, lack of natural light, excessive lighting and contrast levels, and glare can all have a negative impact on the health and productivity of building occupants. Some of these negative impacts include eyestrain, eye irritation, blurred vision, burning eyes, headaches, and depression. Many of the features in the Light Concept help minimize the negative impacts of lighting on the human body's circadian rhythm.

BY THE NUMBERS

- *The Institute of Medicine (Health and Medicine Division) reports that about 50 to 70 million adults in the United States suffer from a chronic sleep or wakefulness disorder.*

Circadian Rhythms

The human body responds to external cues or physical stimuli, such as light, temperature, and eating behaviors, called zeitgebers. Zeitgebers align physiological functions to the 24-hour solar day and influence the circadian rhythm. The *circadian rhythm* is the internal clock that keeps the body's hormones and bodily processes on a roughly 24-hour cycle, even in continuous darkness. The circadian rhythm can influence sleep/wake cycles, hormone release, body temperature, and metabolism, all of which regulate multiple physiological processes such as digestion and sleep. Many health risks, such as diabetes, obesity, depression, heart attack, hypertension, and stroke, increase when these physiological processes are disrupted. Circadian rhythms are heavily influenced by the light that enters the eye and the photoreceptive cells that pick it up.

Photoreceptive Cells. Light that enters the eye is picked up by three types of photoreceptive cells at the back of the eye: rods, cones, and intrinsically photosensitive retinal ganglion cells (ipRGCs). **See Figure 8-7.** The function of these cells is to absorb light and transmit information in the form of electrochemical signals to different parts of the brain. Rods, cones, and ipRGCs each have their own function and peak sensitivity.

The first two types of photoreceptive cells, rods and cones, facilitate vision. Rods facilitate peripheral vision and vision in dim lighting conditions, but they offer low visual acuity and color perception. Rods have a peak sensitivity to green-blue light (498 nm). Conversely, cones facilitate daytime vision, higher visual acuity, and color perception. Cones have a peak sensitivity to green-yellow light (555 nm). **See Figure 8-8.**

However, it is the ipRGCs that mostly facilitate the synchronization of the circadian rhythms. These cells demonstrate peak sensitivity to teal-blue light (≈480 nm). The ipRGCs in the eye contain melanopsin, which is a photopigment that aids in maintaining circadian rhythms. Exposure to excessive amounts of light in the 480 nm wavelength later in the day can alter people's circadian rhythms and cause a disruption in their biological cycles.

The difference between the visual lux and the amount of light that will affect circadian rhythm is known as equivalent melanopic lux. *Equivalent melanopic lux (EML)* is a measure of light that is used to quantify how much a light source will stimulate the light response of melanopsin.

PHOTORECEPTIVE CELLS

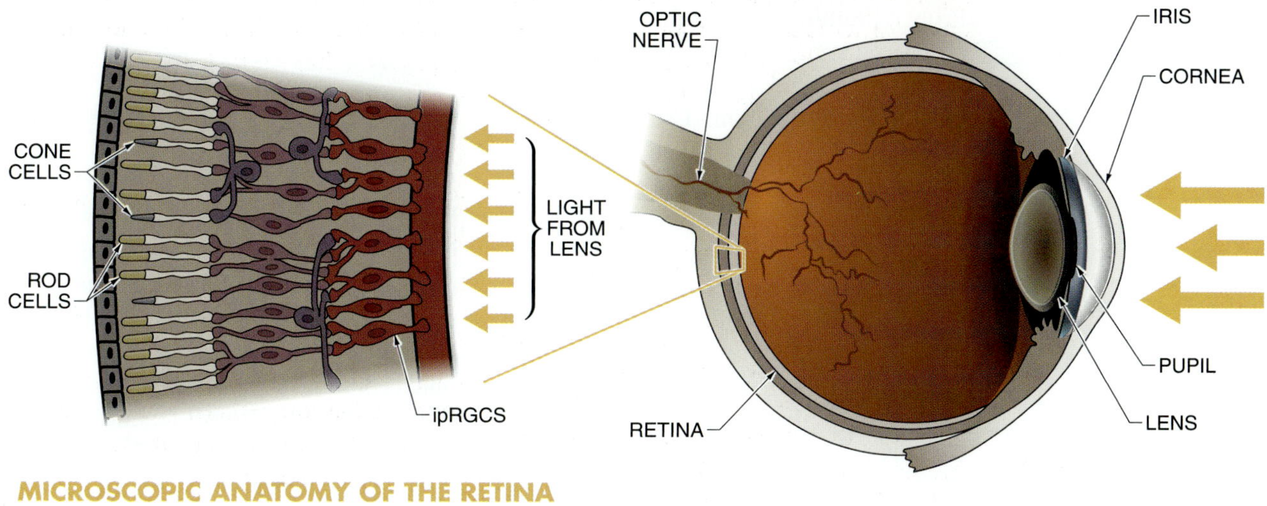

Figure 8-7. The human eye contains three types of photoreceptive cells: rods, cones, and intrinsically photosensitive retinal ganglion cells (ipRGCs).

PHOTORECEPTIVE CELL SENSITIVITY

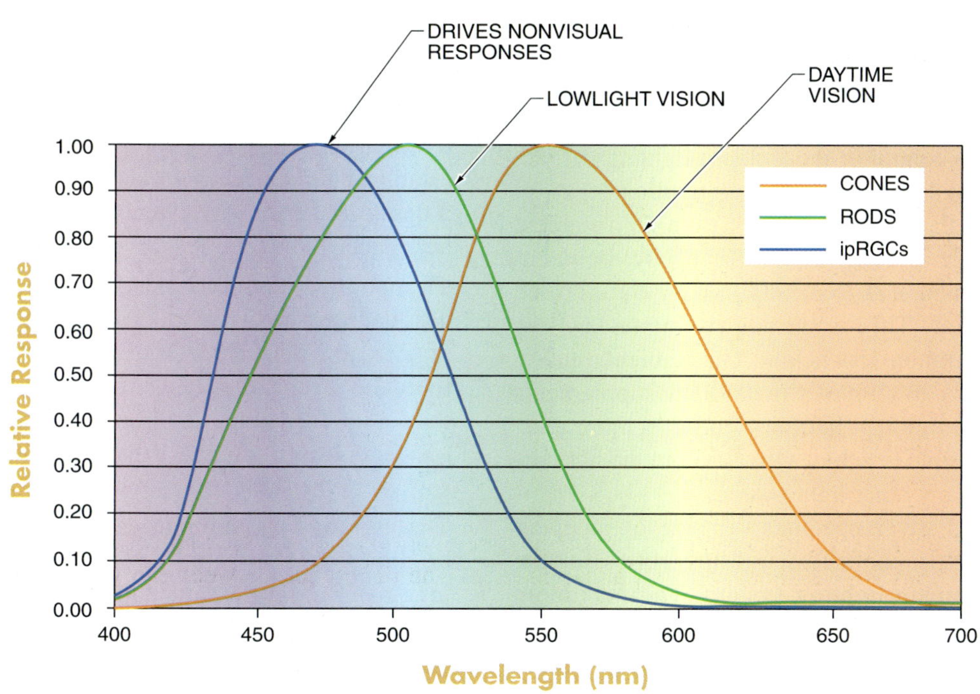

Figure 8-8. Rods, cones, and ipRGCs have different peak sensitivities to particular wavelengths of light.

FEATURE 53, VISUAL LIGHTING DESIGN

Proper illumination levels are needed in order to maintain the proper environment for visual acuity. *Visual acuity* is the clarity or sharpness of vision. The type of work being performed in a space is one of the primary factors in determining the required illumination level. The more detailed work that needs to be performed, the greater the illumination that is needed.

Feature 53, Visual Lighting Design, is a precondition for the New and Existing Interiors and New and Existing Buildings project types that establishes light levels for basic visual performance. Ambient light levels must be sufficient to allow for the completion of most tasks. Task lighting that provides sufficient additional illumination must be available upon request if the ambient lighting does not meet the illumination levels set in the feature. A narrative for brightness management is also required.

Part 1: Visual Acuity for Focus

Part 1: Visual Acuity for Focus of Feature 53 includes requirements for all workstations or desks in a building. The requirements of Part 1a–c pertain to the levels of ambient light, the control of the ambient lighting system, and the levels of any task lighting.

Part 1a requires that a minimum ambient light intensity of 215 lux (20 fc) be maintained at 0.76 m (30″) above the finished floor. This is the approximate level of common desk height. The ambient lighting may be dimmed if daylight supplements the lighting, but the electric lighting must be able to achieve this level in the absence of any other light source.

Part 1b specifies the size of independently controlled lighting zones for the ambient lighting system. The maximum size of an ambient lighting zone that meets the requirements is 46.5 m^2 (500 ft^2) or 20% of the open floor space, whichever is larger.

Part 1c specifies the intensity of the task lighting if ambient lighting does not provide sufficient illumination. If the ambient light is below 300 lux (28 fc), then task lighting must be used to provide 300 lux to 500 lux (28 fc to 46 fc) at work surfaces when needed.

Part 2: Brightness Management Strategies

Part 2: Brightness Management Strategies of Feature 53 addresses the brightness contrasts in the occupied spaces of a building. Brightness contrast is important because while some difference in brightness is helpful in creating visual hierarchies in a space, too much contrast may cause glare or eyestrain. Part 2 requires a project team to provide a narrative describing strategies to maintain luminance balance in spaces. The narrative must take into account at least two of the following:

- The narrative should consider the maximum brightness contrasts between main rooms and ancillary spaces, such as corridors and stairwells, if present. For example, it may be established that a main room cannot exhibit 10 times greater or lesser luminance than an ancillary space.
- The narrative should consider the maximum brightness contrasts between task surfaces and immediately adjacent surfaces, including adjacent visual display terminal screens. For example, it may be established that a surface cannot exhibit 3 times greater or lesser luminance than an adjacent surface.
- The narrative should consider the brightness contrasts between task surfaces and remote, nonadjacent surfaces in the same room. For example, it may be established that a surface cannot exhibit 10 times greater or lesser luminance than a remote surface in the same room.
- The narrative should consider the way brightness is distributed across ceilings in a given room that maintains lighting variety but avoids both dark spots or excessively bright spots. For example, it may be established that one part of a ceiling cannot exhibit 10 times greater or lesser luminance than another part of the ceiling in the same room.

FEATURE 54, CIRCADIAN LIGHTING DESIGN

Feature 54, Circadian Lighting Design, is a precondition for the New and Existing Interiors and New and Existing Buildings project types. This feature promotes strategies that encourage healthy circadian rhythms. The light levels that can affect a person's physiology by influencing circadian rhythms are measured in equivalent melanopic lux (EML). During performance verification, EML is measured on the vertical plane at the average eye level of the building occupants.

Part 1: Melanopic Light Intensity for Work Areas

Part 1: Melanopic Light Intensity for Work Areas of Feature 54 requires light models or light calculations demonstrating that one of two options is met. Part 1a requires light models or light calculations that show that at least 200 EML is present at 75% or more of workstations, measured on the vertical plane facing forward, 1.2 m (4′) above the finished floor. This height represents the average eye-level of a person when seated. This light level may incorporate daylight and must be present every day of the year at least from 9:00 AM to 1:00 PM.

Part 1b requires that electric lights for all workstations provide a maintained illuminance on the vertical plane facing forward (simulating the occupant's view) of 150 EML or greater. The design EML level can be determined by using Table L1: Melanopic Ratio or Table L2: Melanopic and Visual Response located in Appendix C of the WELL Building Standard.

- Table L1 employs melanopic ratios that are dependent on the CCT and the type of light source. The equation $EML = L$ (lux) × R (melanopic ratio) is provided to find the EML level. For example, an LED lamp with a CCT of 4000K has a melanopic ratio of 0.76. If the same lamp has 300 lux of visual brightness, then the equivalent melanopic lux would be 228 (300 lux × 0.76 = 228 EML).
- Table L2 uses the light output of a lamp provided by the manufacturer or a spectrometer. The light output at each wavelength is multiplied by the melanopic and visual curves for each wavelength. The totals are determined, and the total melanopic response is divided by the total visual response. This result is multiplied by 1.218.

FEATURE 55, ELECTRIC LIGHT GLARE CONTROL

Glare is the excessive brightness, excessive brightness contrast, and excessive quantity of light from a light source. Glare interferes with visual perception due to large light differences within the visual field. The first part of Feature 55, Electric Light Glare Control, is a precondition for the Existing Interiors and New and Existing Buildings project types, but the second part is a precondition for all three types. The goal of this feature is to reduce the amount of glare that might be produced by interior light sources.

Shielding is used on interior light fixtures to reduce glare and evenly distribute the quantity of light.

Part 1: Luminaire Shielding

To reduce the amount of glare caused by interior light sources, shielding is required for the light fixtures. The angle of the shielding is dependent on the luminance of the lamp (in cd/m^2). The greater the luminance, the greater the shielding angle required. **See Figure 8-9.**

SHIELDING ANGLES

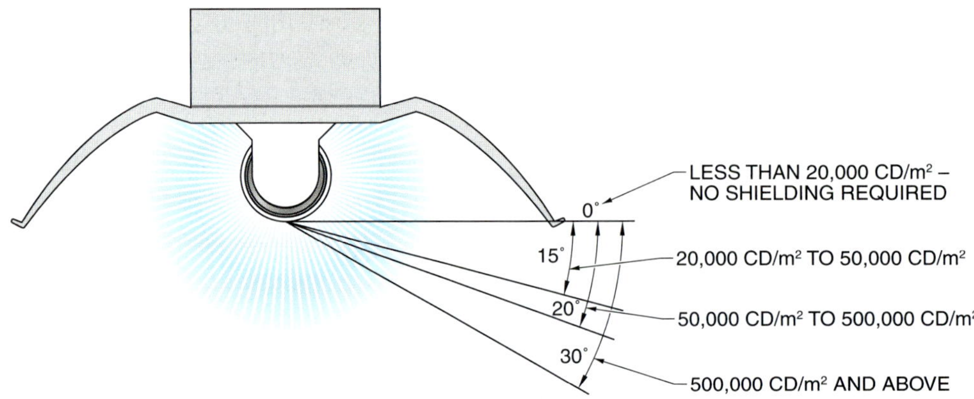

Figure 8-9. Shielding angles are dependent on the level of luminance of the lamp. The greater the luminance, the greater the angle required.

Part 1: Luminaire Shielding of Feature 55 requires that minimum shielding angles be observed in regularly occupied spaces depending on the following luminance values of the lamps:

- less than 20,000 cd/m^2: no shielding required
- 20,000 cd/m^2 to 50,000 cd/m^2: a shielding angle of 15°
- 50,000 cd/m^2 to 500,000 cd/m^2: a shielding angle 20°
- 500,000 cd/m^2 and above: a shielding angle 30°

Part 2: Glare Minimization

Part 2: Glare Minimization of Feature 55 applies to all workstations, desks, and other seating areas. Luminaires that are located more than 53° above the center of view of work spaces (meaning they are placed almost or directly above the work space) can cause excessive glare and discomfort for the people in the work spaces. To limit excessive glare and discomfort, Part 2: Glare Minimization of Feature 55 requires that any light source more than 53° above the center of view must have luminance values of less than 8000 cd/m^2. **See Figure 8-10.**

GLARE MINIMIZATION

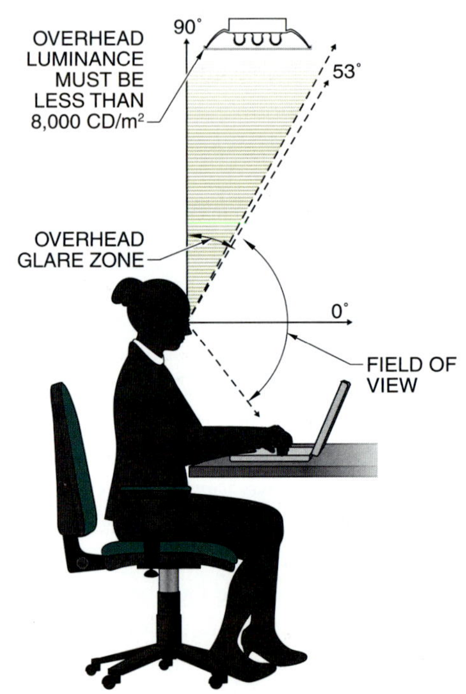

Figure 8-10. Part 2: Glare Minimization of Feature 55, Electric Light Glare Control, requires that any lighting more than 53° above the center of view have luminance values less than 8000 cd/m^2.

FEATURE 56, SOLAR GLARE CONTROL

Feature 56, Solar Glare Control, is an optimization for Core and Shell projects, but it is a precondition for the New and Existing Interiors and New and Existing Buildings project types. The goal of this feature is to limit the effects of glare for building occupants through the installation or integration in the building design of manual or automatic shading solutions.

This feature is split into requirements for the shading of view windows and the shading of daylight windows. WELL differentiates between these two types of windows by considering glazing below 2.1 m (7′) to be a view window and glazing above 2.1 m (7′) to be a daylight window. **See Figure 8-11.** A *view window* is glazing that is designed to allow building occupants clear views of the outside. A *daylight window* is glazing that is specifically designed and located to allow for deeper penetration of daylight into the interior spaces of a building. Often, view windows and daylight windows are specified as the same glazing, but they are treated differently in regards to daylighting control requirements.

VIEW AND DAYLIGHT WINDOW GLAZING

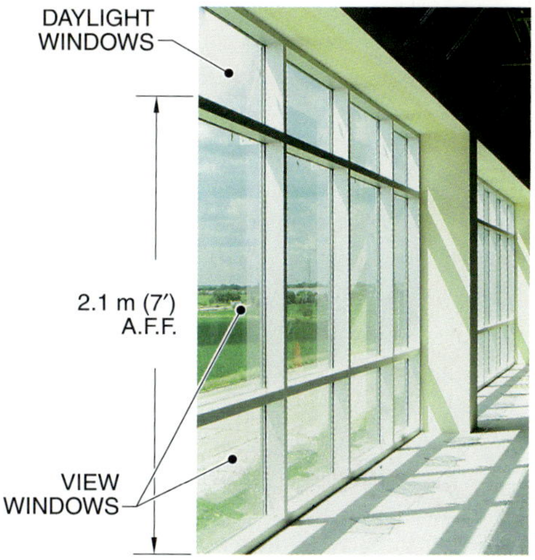

Figure 8-11. WELL differentiates between view and daylight windows by considering glazing below 2.1 m (7′) to be a view window and glazing above 2.1 m (7′) to be a daylight window.

Part 1: View Window Shading

Part 1: View Window Shading of Feature 56 requires that at least one of three options be implemented for glazing less than 2.1 m (7′) above the floor in regularly occupied spaces. Part 1a includes the option of interior window shading or blinds that are either manually controlled or automatically set to prevent glare. Part 1b includes the option of exterior shading systems that are set to prevent glare. Part 1c includes the option of variable opacity glazing that reduces transmissivity by 90% or more.

Variable opacity glazing, also known as smart glass, is glazing that has the ability to control the amount of light passing through it. One commonly used type of variable opacity glazing is electrochromic glass, which has an ultrathin coating that changes color when a low-voltage electric current is applied. **See Figure 8-12.**

ELECTROCHROMIC GLASS

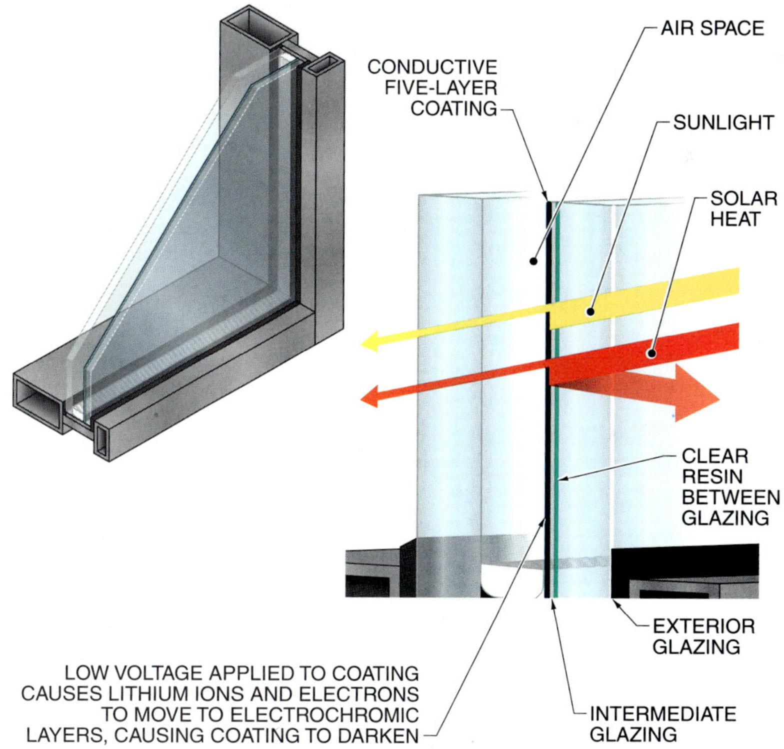

Figure 8-12. Electrochromic glass has an ultrathin coating that changes color when a small electric current is applied.

Manually operated interior window blinds allow building occupants to control the amount of natural daylight at workstations and reduce glare.

Part 2: Daylight Management

Part 2: Daylight Management of Feature 56 requires that at least one of five options be implemented for glazing more than 2.1 m (7′) above the floor in regularly occupied spaces. The first two options are shading solutions that can have manual or automatic controls. The other options are static design elements. **See Figure 8-13.** The options for Part 2 include the following:

- interior window shading or blinds
- external shading systems
- interior light shelves
- a film of micromirrors on the windows that reflect light toward the ceiling
- variable opacity glazing

FEATURE 57, LOW-GLARE WORKSTATION DESIGN

Feature 57, Low-Glare Workstation Design, is an optimization for the New and Existing Interiors and New and Existing Buildings project types. The goal of this feature is to help reduce the amount of glare from bright light sources, specifically at workstations. Bright light sources, like those from windows or overhead luminaires, can cause eye fatigue, headaches, and productivity losses. Building occupants may also experience other physical discomforts since they may continually change their orientation to the computer screen to avoid the glare.

Part 1: Glare Avoidance

Part 1: Glare Avoidance of Feature 57 has two requirements that must be met. Part 1a requires that all computer screens at desks that are located within 4.5 m (15′) of view windows must be able to be oriented within a 20° angle perpendicular to the plane of the nearest window. **See Figure 8-14.** Part 1b requires that overhead luminaires not be aimed directly at computer screens. This can be accomplished by locating workstations between overhead lights or by using diffusers or defectors.

DAYLIGHT MANAGEMENT

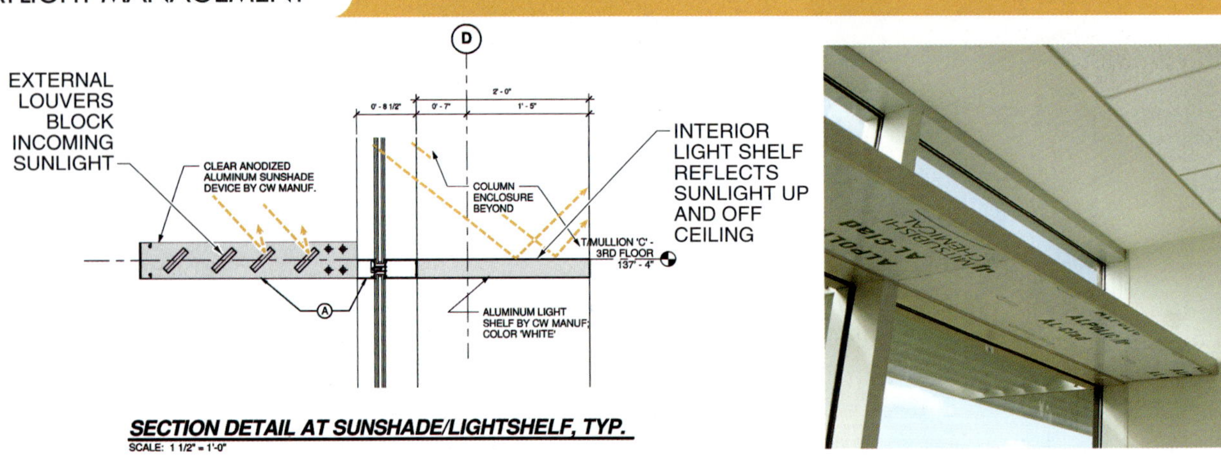

Figure 8-13. External shading systems and interior light shelves are two strategies for Part 2: Daylight Management of Feature 56, Solar Glare Control.

GLARE AVOIDANCE

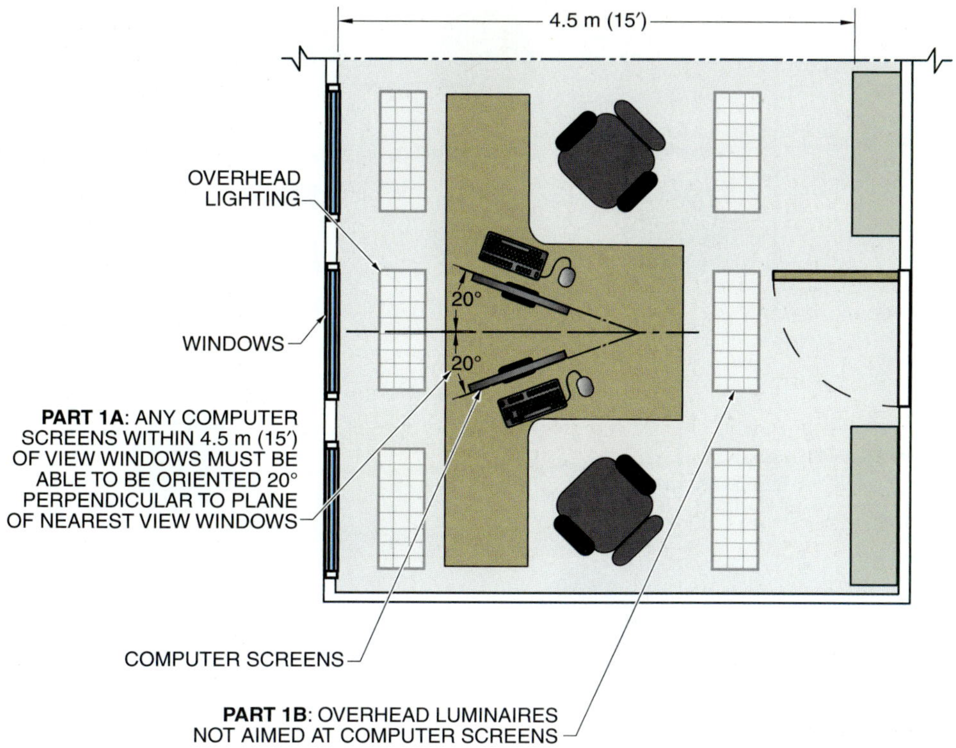

Figure 8-14. The first requirement for Part 1: Glare Avoidance of Feature 57, Low-Glare Workstation Design, is that all computer screens at desks that are located within 4.5 m (15′) of view windows must be able to be oriented within a 20° angle perpendicular to the plane of the nearest window.

FEATURE 58, COLOR QUALITY

Feature 58, Color Quality, is an optimization for the New and Existing Interiors and New and Existing Buildings project types. Color quality is important because inadequate color rendering can have a negative effect on building occupant comfort. Color quality is measured using the color rendering index. The *color rendering index (CRI)* is an index that features a comparison of the appearance of 8 to 14 colors under a light source in comparison to a blackbody source of the same color temperature. CRI is measured using a palate of reference colors called R1 through R14. The color rendering index average (CRI Ra) is a value on a scale of 1 to 100 based on R1 through R8. **See Figure 8-15.**

WELLNESS FACT

A blackbody is an ideal surface or object that absorbs all radiation directed toward it and reflects none in return. These theoretical nonreflective bodies are useful for comparing the luminosity of other surfaces or objects.

Part 1: Color Rendering Index

Part 1: Color Rendering Index of Feature 58 applies to all electric lights, except for decorative, emergency, or special-purpose lighting. Part 1a requires that the CRI Ra be 80 or higher. Part 1b requires that the CRI R9 value be 50 or higher. R9 represents the red tones and allows a more accurate representation of actual color.

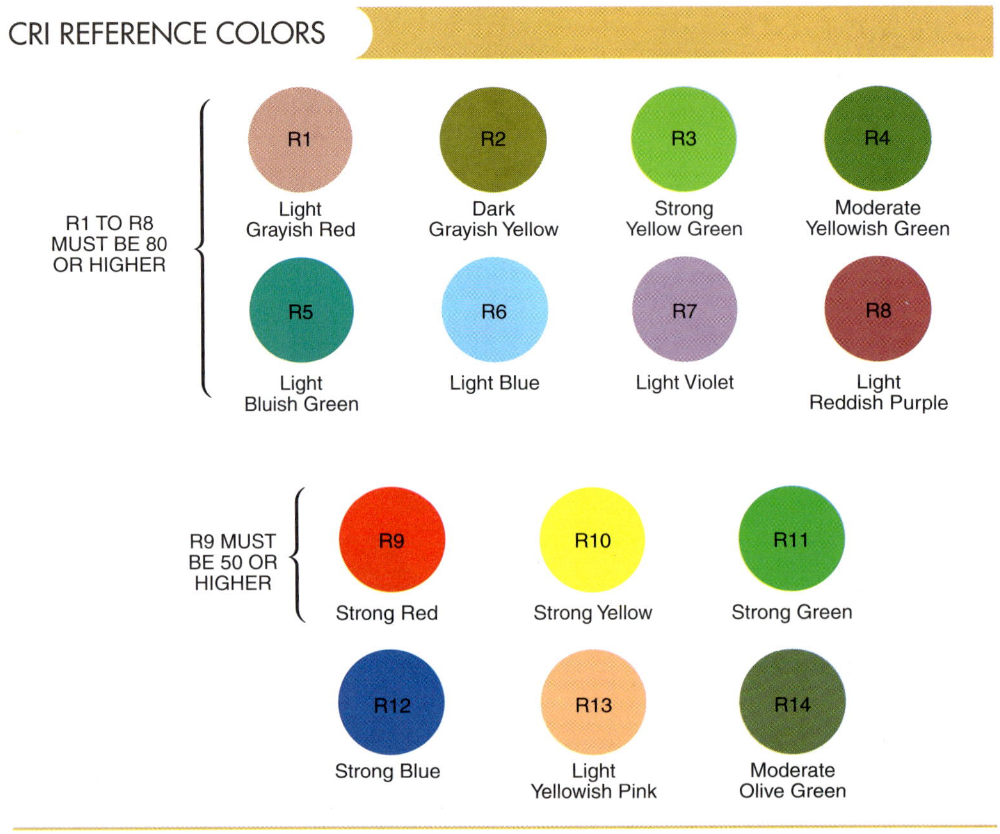

Figure 8-15. The CRI reference colors are R1 through R14, and the color rendering index average (CRI Ra) is a value on a scale of 1 to 100 based on R1 through R8.

FEATURE 59, SURFACE DESIGN

Feature 59, Surface Design, is an optimization for the New and Existing Interiors and New and Existing Buildings project types. Lighting in the built environment takes two forms: direct light and reflected light. Direct light comes from luminaires and incoming sunlight. Reflected light is the direct light that has reflected off the ceilings, vertical surfaces such as walls, and interior furnishings. The goal of this feature is to increase the amount of light for better visual acuity and regulation of the circadian rhythms of building occupants.

The higher the reflective quality of the interior surfaces, the greater the amount of reflected light that will result. The reflective quality of surfaces is rated using a method called light reflectance value. *Light reflectance value (LRV)* is a rating of how much usable light is reflected or absorbed by a given surface, with values ranging from 0 (0%) to 1 (100%). An LRV of 0 represents absolute black, while an LRV of 1 represents totally reflective white. **See Figure 8-16.**

Part 1: Working and Learning Area Surface Reflectivity

Part 1: Working and Learning Area Surface Reflectivity of Feature 59 requires that three sets of surfaces in regularly occupied spaces meet minimum LRVs. The three sets of surfaces and their minimum LRVs are ceilings (0.8 for 80% of surface area), vertical surfaces (0.7 for 50% of surface area), and furniture systems (0.5 for 50% of surface area). **See Figure 8-17.**

LIGHT REFLECTANCE VALUES (LRV)

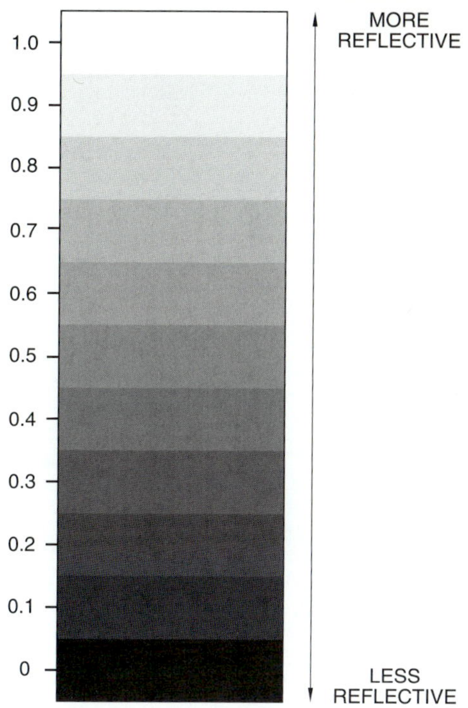

Figure 8-16. The LRV is a rating of how much usable light is reflected or absorbed by a given surface, with values ranging from 0 (0%) to 1 (100%).

FEATURE 60, AUTOMATED SHADING AND DIMMING CONTROLS

Feature 60, Automated Shading and Dimming Controls, is an optimization for the New and Existing Interiors and New and Existing Buildings project types. The goals of this feature are to control the amount of light coming into a space by allowing window shades to be automatically raised and lowered as required and to automatically dim or shut off the lighting in a space when it is unoccupied or in response to daylight levels.

Part 1: Automated Sunlight Control

Part 1: Automated Sunlight Control of Feature 60 applies to windows larger than 0.55 m^2 (6 ft^2). Part 1 requires that all applicable windows have automatic shading devices for reducing glare at workstations and other seating areas. An automated shading device is part of a sensor-controlled sunlight control system that automatically raises and lowers the shades at predetermined levels of sunlight. **See Figure 8-18.** These systems can be fully integrated into a building automation system or can be stand-alone systems.

WORKING AND LEARNING AREA SURFACE REFLECTIVITY

Figure 8-17. Part 1: Working and Learning Area Surface Reflectivity of Feature 59, Surface Design, requires that the ceilings, vertical surfaces, and furniture systems in regularly occupied spaces meet minimum LRV percentages.

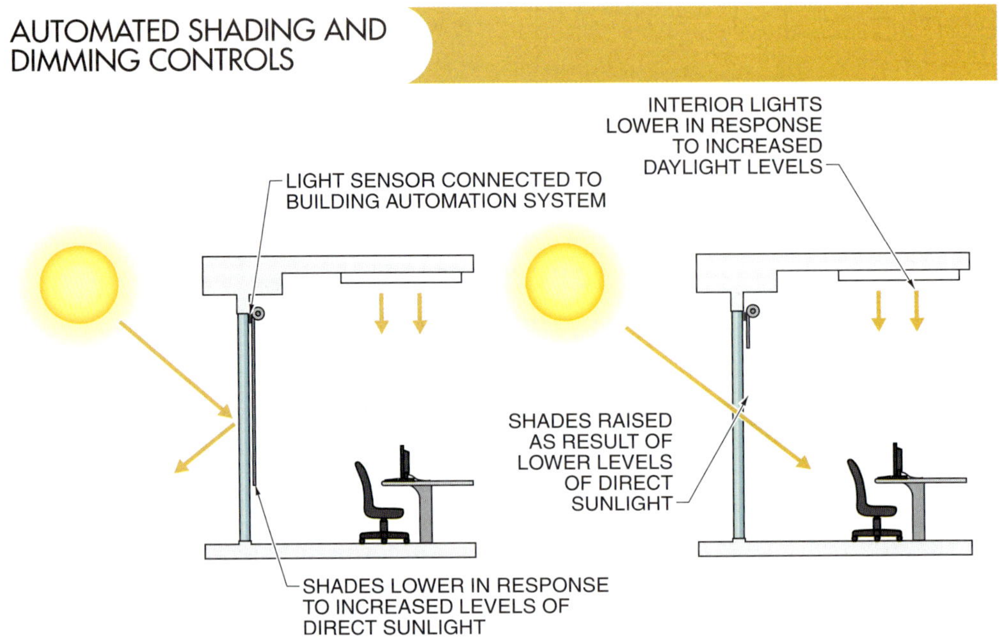

Figure 8-18. Feature 60, Automated Shading and Dimming Controls, requires the use of automated shade devices to control the amount of daylight coming into a space and programmable occupancy sensors to limit artificial light output.

Part 2: Responsive Light Control

Part 2: Responsive Lighting Control of Feature 60 requires that all lighting, except decorative fixtures, in all major work areas have automatic control sensors. Part 2a requires the lighting to automatically dim to 20% or less, or shut off, when the zone is unoccupied. Part 2b requires that the fixtures in the space be programmed to dim continuously as daylight levels increase.

FEATURE 61, RIGHT TO LIGHT

The first part of Feature 61, Right to Light, is an optimization for all three project types, but the second part is an optimization only for the New and Existing Interiors and New and Existing Buildings types. This feature requires that a majority of the regularly occupied spaces have access to sunlight and that any workstations within a given distance have access to views of the exterior. Access to natural light and outside views are sustainable design features that have synergy with many other green building certifications, such as LEED.

Part 1: Lease Depth

Part 1: Lease Depth of Feature 61 requires that 75% of the area of all regularly occupied spaces be within 7.5 m (25′) of view windows. **See Figure 8-19.** Often, this means that the majority of the regularly occupied spaces must be designed to be near the perimeter of the building where view windows are generally located or in another area where windows provide a view of the exterior.

Part 2: Window Access

Part 2: Window Access of Feature 61 requires that workstations be within a given distance of an atrium or allow views of the exterior. View windows and atriums allow natural daylight to provide the primary source of lighting. Part 2a requires that 75% of workstations be within 7.5 m (25′) of a view window or atrium. Part 2b expands this requirement up to 95% of workstations but moves the threshold back to 12.5 m (41′). **See Figure 8-20.**

RIGHT TO LIGHT: LEASE DEPTH

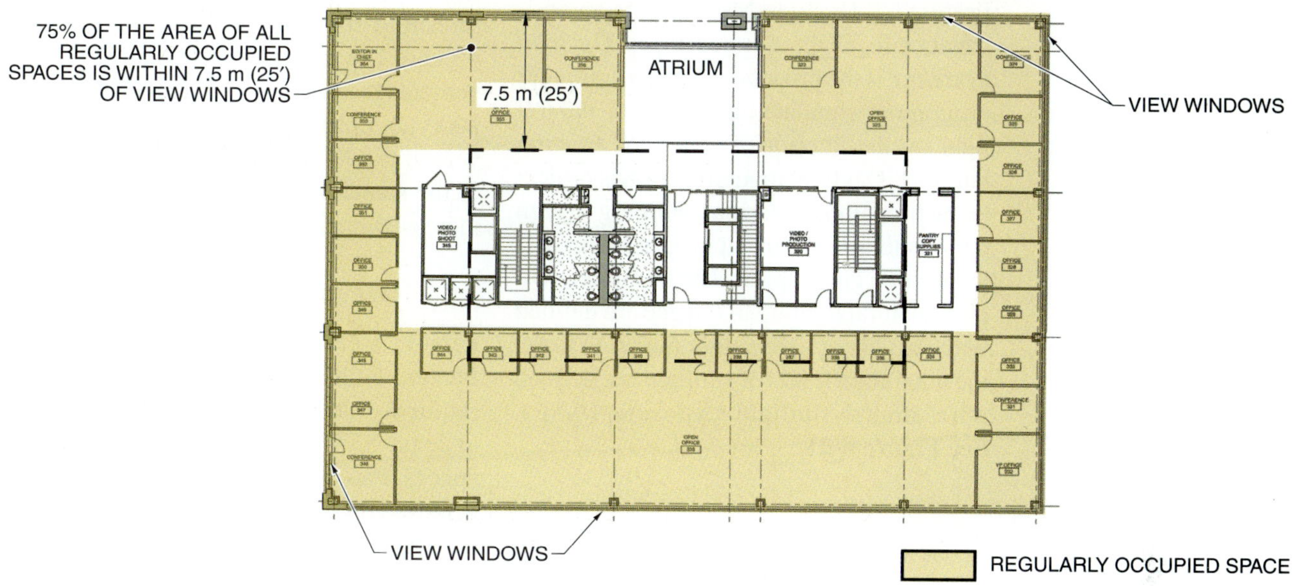

Figure 8-19. Part 1: Lease Depth of Feature 61, Right to Light, requires that 75% of the area of all regularly occupied spaces be within 7.5 m (25′) of view windows.

RIGHT TO LIGHT: WINDOW ACCESS

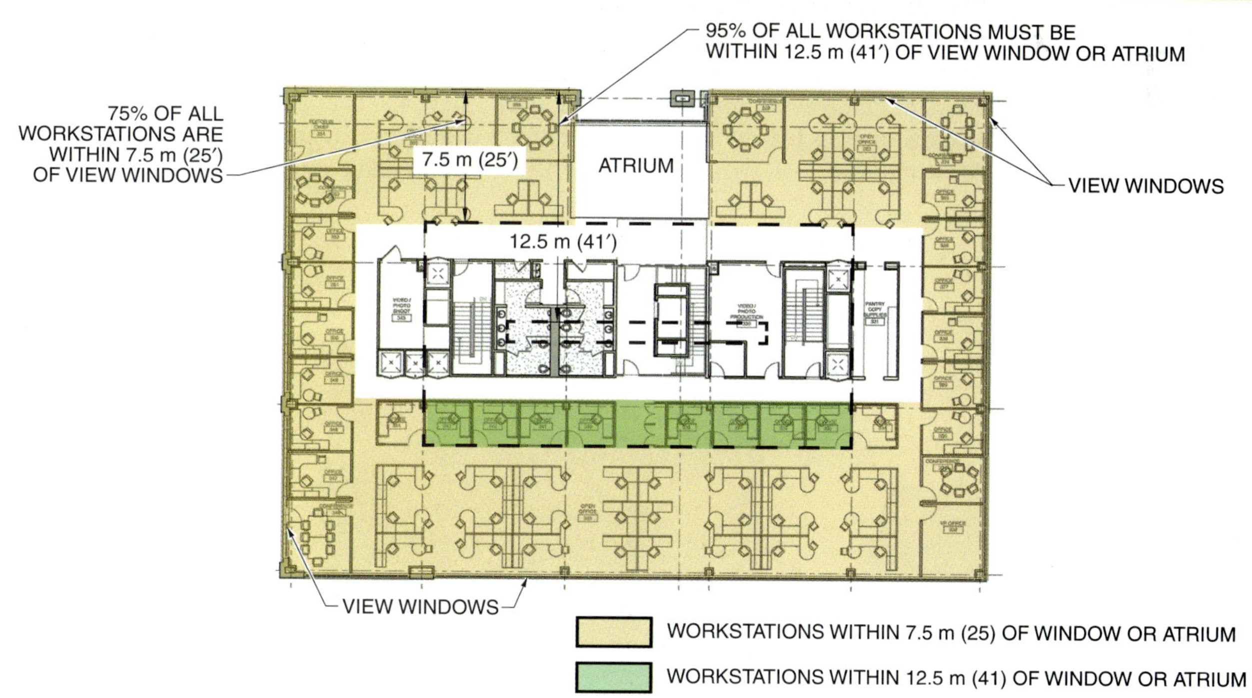

Figure 8-20. Part 2: Window Access of Feature 61, Right to Light, requires that workstations be within a given distance of an atrium or views of the exterior.

FEATURE 62, DAYLIGHT MODELING

Feature 62, Daylight Modeling, is an optimization for all three project types. *Daylight modeling* is the process of using a computer program to simulate the effects of natural light on the interior of a new or existing building. Daylight modeling allows project teams to compare various building design options to achieve the appropriate lighting levels for a given space. The minimum levels of natural lighting in a space are defined with spatial daylight autonomy (sDA), and the maximum levels of sunlight are defined with annual sunlight exposure (ASE). **See Figure 8-21.**

WELLNESS FACT

The daylight metrics sDA and ASE are complementary to ensure proper levels of sunlight enter a space without causing thermal discomfort.

Spatial daylight autonomy (sDA) is a percentage of floor space in which a minimum light level can be met completely for some proportion of regular operating hours by natural light. The criteria of $sDA_{300,50\%}$ requires that a minimum light level of 300 lux (28 fc) be achieved for a given area for 50% of the analysis period.

Annual sunlight exposure (ASE) is a percentage of space in which the light level from direct sun alone exceeds a predefined threshold for some quantity of hours in a year. The criteria of $ASE_{1000,250}$ is the percentage of space that receives at least 1000 lux (93 fc) for at least 250 hours per year. If the ASE value is over what is given in the standard, then there is a chance for increased glare and thermal discomfort.

Part 1: Healthy Sunlight Exposure

Part 1: Healthy Sunlight Exposure of Feature 62 has two requirements. Part 1a requires that $sDA_{300,50\%}$ be achieved for at least 55% of the regularly occupied spaces. Part 1b requires that $ASE_{1000,250}$ be achieved for no more than 10% of regularly occupied spaces.

FEATURE 63, DAYLIGHTING FENESTRATION

Feature 63, Daylighting Fenestration, is an optimization for all three project types. The goal of this feature is to optimize a project's windows in order to increase the amount of daylight available for the building spaces while limiting the amount of glare and heat gain. This feature addresses window sizes and locations, light transmittance, and color transmittance.

DAYLIGHT MODELING

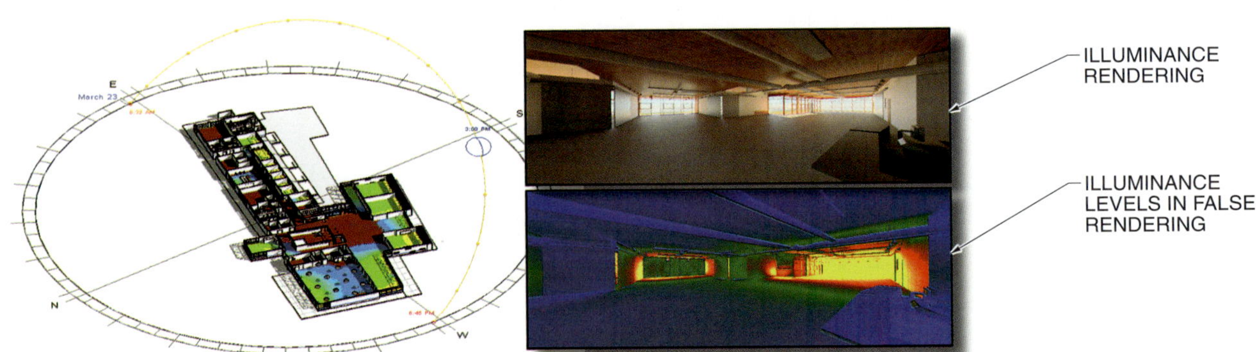

Autodesk, Inc.

Figure 8-21. Daylight modeling is the process of using a computer program to simulate the effects of natural light on the interior of a new or existing building.

Part 1: Window Sizes for Working and Learning Spaces

Part 1: Window Sizes for Working and Learning Spaces of Feature 63 has two conditions that must be met on facades along regularly occupied spaces. Part 1a requires that the window-wall ratio on the external elevations of a building be between 20% and 60%. Spandrel glass, which is glazing designed to hide construction features between floors, is not included as part of the window calculation, but rather as an opaque wall. **See Figure 8-22.** If the ratio is greater than 40%, then strategies, such as internal and external shading systems, must be in place to control unwanted heat gain and glare. Part 1b requires that 40% to 60% of the window area be at least 2.1 m (7′) off the floor, which means it must be a daylight window.

Part 2: Window Transmittance in Working and Learning Areas

Part 2: Window Transmittance in Working and Learning Areas of Feature 63 requires that visible transmittance (VT) conditions are met for all nondecorative glazing. *Visible transmittance (VT)* is the amount of light in the visible portion of the spectrum (roughly 400 nm to 700 nm) that passes through a glazing material. VT is measured as a percentage or number from 0 to 1. The higher the percentage, the greater the amount of visible light that is allowed through.

WELLNESS FACT

The glazing for view windows is typically located at the average eye level of building occupants to allow views of the outdoors, which can positively affect the mood and health levels of occupants. The glazing for daylight windows is located above average eye level and primarily serves to allow the penetration of sunlight into a space.

Part 2a requires that all nondecorative glazing, excluding skylights, located above 2.1 m (7′) have a VT of 60% or more. This value allows for an increased amount of daylight in the space. **See Figure 8-23.** Part 2b requires that all nondecorative glazing located below 2.1 m (7′) have a VT of 50% or more. A lower value for the lower glazing reduces the amount of glare that could affect the building occupants.

WINDOW SIZES FOR WORKING AND LEARNING SPACES

Figure 8-22. Part 1: Window Sizes for Working and Learning Spaces of Feature 63, Daylighting Fenestration, requires that the window-wall ratio on the external elevations be between 20% and 60% and that 40% to 60% of the window area be at least 2.1 m (7′) off the floor.

VISIBLE TRANSMITTANCE

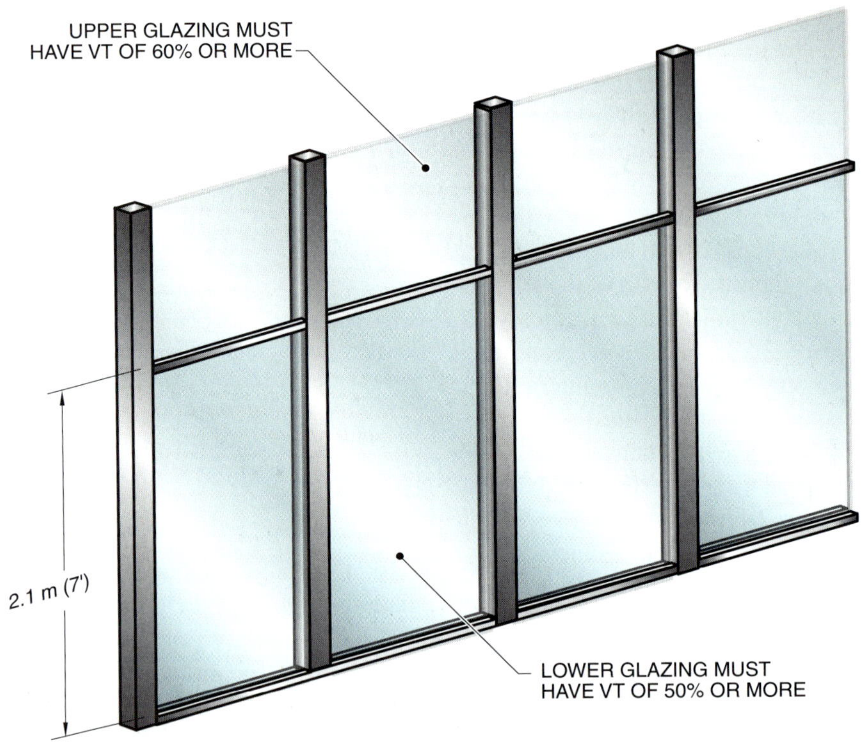

Figure 8-23. Part 2: Window Transmittance in Working and Learning Areas of Feature 63, Daylighting Fenestration, requires that all nondecorative glazing located above 2.1 m (7′) have a VT of 60% or more to allow for an increased amount of daylight in the space and that all nondecorative glazing located below 2.1 m (7′) have a VT of 50% or more.

Part 3: Uniform Color Transmittance

Part 3: Uniform Color Transmittance of Feature 63 specifies the wavelengths of visible light that are allowed through daylight windows. Tinted glass, glass color, or selective glaze coatings that limit the amounts of UV light and IR light into the building may also affect the transmission of visible light.

Part 3 requires that the VT of wavelengths between 400 nm and 650 nm not vary by more than a factor of 2 for daylight glazing. This means that the highest VT for any wavelength between 400 nm and 650 nm cannot be more than 2 times the lowest VT for any wavelength between 400 nm and 650 nm. For example, if the lowest VT for daylight windows is 40% at 460 nm and the highest is 70% at 600 nm, the glazing would be acceptable. However, if the highest VT were 90% at 600 nm while the lowest VT is still 40% at 460 nm, then the glazing would not be acceptable.

KEY TERMS AND DEFINITIONS

accent lighting: Lighting that is used to highlight architectural or design features of a space or to add visual interest to a space. Also known as decorative lighting.

ambient lighting: The main source of nonspecific illumination in a space. Also known as general lighting.

annual sunlight exposure (ASE): A percentage of space in which the light level from direct sun alone exceeds a predefined threshold for some quantity of hours in a year.

arc tube: The light-producing element of an HID lamp.

circadian rhythm: The internal clock that keeps the body's hormones and bodily processes on a roughly 24-hour cycle, even in continuous darkness.

color rendering index (CRI): A comparison of the appearance of 8 to 14 colors under a light source in question to a blackbody source of the same color temperature.

correlated color temperature (CCT): The spectral distribution of electromagnetic radiation of a blackbody at a given temperature, measured in degrees Kelvin (K).

daylight modeling: The process of using a computer program to simulate the effects of natural light on the interior of a new or existing building.

daylight window: Glazing that is specifically designed and located to allow for deeper penetration of daylight into the interior spaces of a building.

electromagnetic radiation: Energy in the form of electromagnetic waves, including gamma rays, X-rays, ultraviolet (UV) light, visible light, infrared (IR) light, microwave radiation, and radio waves.

electromagnetic spectrum: The range of all types of electromagnetic radiation.

equivalent melanopic lux (EML): A measure of light that is used to quantify how much a light source will stimulate the light response of melanopsin.

fluorescent lamp: A low-pressure discharge lamp in which ionization of mercury vapor transforms UV light generated by the discharge into visible light.

glare: The excessive brightness, excessive brightness contrast, and excessive quantity of light from a light source.

high-intensity discharge (HID) lamp: A lamp that produces light from an arc tube.

illuminance: The amount of light incident on a given surface measured in lux or footcandles.

incandescent lamp: A lamp that produces light by the flow of current through a tungsten filament inside a sealed glass bulb sometimes filled with a gas.

lamp: An output component that converts electrical energy into light.

light-emitting diode (LED) lamp: A lamp with semiconductor devices that produce visible light when an electrical current passes through them.

light reflectance value (LRV): A rating of how much usable light is reflected or absorbed by a given surface, with values ranging from 0 (0%) to 1 (100%).

luminaire: A complete lighting unit that includes the components that distribute light, position and protect the lamps, and provide connection to the power supply.

luminance: A measurement of how bright a surface or light source will appear to the eye, measured in candela per square meter (cd/m^2) or footlamberts.

KEY TERMS AND DEFINITIONS *(continued)*

luminous flux: The total luminous output of a light source, measured in lumens and weighted to the visual sensitivity of the human eye.

luminous intensity: Radiant power weighted to human vision that describes light emitted by a source in a particular direction.

spatial daylight autonomy (sDA): A percentage of floor space in which a minimum light level can be met completely for some proportion of regular operating hours by natural light.

task lighting: Lighting that is used to illuminate an area in order to allow the performance of a specific function.

variable opacity glazing: Glazing that has the ability to control the amount of light passing through it. Also known as smart glass.

view window: Glazing that is designed to allow building occupants clear views of the outside.

visible transmittance (VT): The amount of light in the visible portion of the spectrum (roughly 400 nm to 700 nm) that passes through a glazing material.

visual acuity: The clarity or sharpness of vision.

wavelength (λ): The distance between two points on a wave in which the wave repeats itself.

CHAPTER REVIEW

Completion

______________________ **1.** The wavelengths of visible light range from ___ nm to ___ nm.

______________________ **2.** The lower the correlated color temperature (CCT) value in degrees Kelvin, the "___" the light is perceived.

______________________ **3.** ___ is the modulation, or change of intensity, of light output that can cause headaches, eyestrain, or eye discomfort.

______________________ **4.** ___ is the total perceived amount of light being given off in all directions.

______________________ **5.** ___ is a measurement of how bright a surface or light source will appear to the eye and must be balanced to avoid glare and visual discomfort.

______________________ **6.** The internal clock that keeps a body's hormones and bodily processes on a roughly 24-hour cycle is called the ___.

______________________ **7.** Intrinsically photosensitive retinal ganglion cells (ipRGCs) demonstrate peak sensitivity to ___ light.

______________________ **8.** The first requirement for Part 1: Visual Acuity of Feature 53, Visual Lighting Design, is that a minimum ambient light intensity of ___ must be maintained at 0.76 m (30″) above the finished floor, or about the height of a common desk.

CHAPTER REVIEW *(continued)*

_______________ **9.** The light levels that can affect a person's physiology by influencing circadian rhythms are measured in ___.

_______________ **10.** Luminaires that are located more than ___° above the center of view of work spaces can cause excessive glare and discomfort for the people in those spaces and must have luminance values of less than 8000 cd/m².

_______________ **11.** One commonly used type of variable opacity glazing is ___, which has an ultrathin coating that changes color when a low-voltage electric current is applied.

_______________ **12.** The goal of Feature 57, ___, is to help reduce the amount of glare from bright light sources, specifically at workstations.

_______________ **13.** Part 1: Color Rendering Index of Feature 58, Color Quality, requires that the CRI Ra be ___ or higher.

_______________ **14.** The goals of Feature 60, ___, are to control the amount of light coming into a space by allowing window shades to be automatically raised and lowered as required and to automatically dim or shut off the lighting in a space when it is unoccupied or in response to daylight levels.

_______________ **15.** Part 1: Lease Depth of Feature 61, Right to Light, requires that ___% of the area of all regularly occupied spaces be within 7.5 m (25′) of view windows.

_______________ **16.** If the window-wall ratio is greater than ___%, then Part 1: Window Sizes for Working and Learning Spaces of Feature 63, Daylighting Fenestration, requires that internal or external shading systems be in place to control unwanted heat gain and glare.

Short Answer

1. List the three classes of lighting that are used in the built environment.

2. What are the bodily and physiological processes influenced by the circadian rhythm?

3. List the three types of photoreceptive cells and their peak sensitivities.

4. List the strategies for maintaining luminaire balance listed in Part 2: Brightness Management Strategies of Feature 53, Visual Lighting Design.

CHAPTER REVIEW *(continued)*

5. What is the equation for determining EML level provided by Table L1: Melanopic Ratio in Appendix C of the WELL Building Standard?

6. How does WELL differentiate between view windows and daylight windows?

7. List the shading solutions and design elements that meet the requirements for Part 2: Daylight Management of Feature 56, Solar Glare Control.

8. List the surfaces and the minimum light reflectance values (LRVs) that must be met to satisfy the requirements for Part 1: Working and Learning Area Surface Reflectivity of Feature 59, Surface Design.

9. Explain daylight modeling and its two metrics.

WELL AP EXAM PRACTICE QUESTIONS

1. What must light models or light calculations show to meet the requirements for Part 1: Melanopic Light Intensity for Work Areas of Feature 54, Circadian Lighting Design?
   A. At least 150 equivalent melanopic lux is present at 50% or more of workstations
   B. At least 200 equivalent melanopic lux is present at 75% or more of workstations
   C. At least 250 equivalent melanopic lux is present at 90% or more of workstations
   D. At least 300 equivalent melanopic lux is present at all workstations

2. What is the only feature in the Light Concept to include a part that is a precondition for all three project types?
   A. Feature 53, Visual Lighting Design
   B. Feature 54, Circadian Lighting Design
   C. Feature 55, Electric Light Glare Control
   D. Feature 56, Solar Glare Control

3. What is the verification method for the on-site assessment required by Feature 57, Low-Glare Workstation Design?
   A. Spot measurement
   B. Visual inspection
   C. Spot check
   D. Performance test

4. What is a good strategy for ensuring that a sufficient amount of light reaches the eyes of building occupants without increasing glare or the usage of electricity?
   A. Choosing surfaces with higher light reflectance values
   B. Choosing surfaces with lower light reflectance values
   C. Redirecting overhead luminaires away from computer screens
   D. Installing windows with low visible transmittance

5. A project has windows that are 0.65 m^2 (7 ft^2) and allow sunlight into occupied areas, causing glare at workstations. What must be installed in order to meet the requirements of Part 1: Automated Sunlight Control of Feature 60, Automated Shading and Dimming Controls?
   A. Interior manual window shading or blinds
   B. Glazing with a VT of 50% or less
   C. Interior light shelves
   D. Automatic shading devices

CHAPTER NINE

FITNESS

Many health problems can be prevented by having a healthy lifestyle. Adequate fitness is an integral part of this lifestyle. Since 90% of a person's time is spent within the built environment, building designers, building owners, and urban planners can promote healthy lifestyles by designing buildings and spaces with occupant physical activity in mind and by implementing programs that encourage physical activity. The goal of the Fitness Concept in the WELL Building Standard is to encourage building occupants to be more physically active and less sedentary.

OBJECTIVES

- Identify the different types of fitness.
- Identify the features in the Fitness Concept as preconditions or optimizations.
- Explain how the features address the health effects that the lack of fitness can have on human body systems.
- Describe the strategies used to increase fitness inside and outside of the workplace.
- Explain the requirements and the benefits of an activity incentive program.
- Identify the equipment that building occupants can use to increase fitness.

KEY TERMS

- active transportation
- cardiorespiratory fitness
- diverse use
- fitness
- pedestrian amenity
- sedentary behavior
- Walk Score®

Learner Resources

ATPeResources.com/QuickLinks
Access Code: 935534

FITNESS CONCEPT

Fitness, as defined by the Centers for Disease Control (CDC), is the ability to carry out daily tasks with vigor and alertness, without undue fatigue, and with ample energy to enjoy leisure-time pursuits and respond to emergencies. Each building occupant is ultimately responsible for adequately maintaining his or her own fitness. However, the WELL Building Standard encourages building owners to increase the accessibility of fitness opportunities and to become more involved in the health and fitness of building occupants.

The Fitness Concept in WELL includes eight features that promote strategies such as interior and exterior design, structured fitness opportunities, and the proper space and equipment for fitness activities. Core and Shell projects must meet one precondition, but four optimizations can be achieved for higher certification. New and Existing Interiors must meet one precondition, but seven optimizations can be pursued for higher certification. New and Existing Buildings must meet two preconditions, but six optimizations can be pursued for higher certification. **See Figure 9-1.**

Employers can promote employee fitness by encouraging alternative methods of transportation such as bicycle commuting and by using subsidies or tax-exempt payroll deductions to incentivize these types of physical activity.

BY THE NUMBERS

- *It is recommended that all healthy adults engage in at least 30 minutes of moderate-intensity aerobic activity 5 days per week and muscle-strengthening activities at least 2 days per week.*

TYPES OF FITNESS

There are three types of fitness, and a complete fitness program should include all three. The three types of fitness are strength and endurance, cardiorespiratory, and flexibility fitness. **See Figure 9-2.** Some physical activities involve one type of fitness while other physical activities involve more than one type.

Strength and Endurance Fitness

Strength and endurance fitness exercises, also known as resistance training, involve building and maintaining the body's muscle mass. Besides increases in strength, strength and endurance fitness exercises can also increase bone density, which may help reduce the risk of osteoporosis. Other benefits include increasing blood flow through the body, lowering blood pressure, and increasing the amount of calories burned throughout the day.

There are many ways to achieve strength and endurance fitness. Activities such as lifting free weights, using weight machines, or performing bodyweight exercises, such as pull-ups, push-ups, or squats, can be used to increase muscle mass. Activities such as boxing or martial arts have the dual benefit of increasing both strength fitness and cardiorespiratory fitness.

Cardiorespiratory Fitness

Cardiorespiratory fitness, also referred to as aerobic fitness, is a component of physical fitness that involves the ability of the body, specifically the heart and lungs, to transport, absorb, and use oxygen during sustained physical activity. Cardiorespiratory fitness increases the strength and efficiency of the heart and lungs. Increased cardiorespiratory fitness can reduce the chances of developing type 2 diabetes, decrease the risk of coronary heart disease, lower blood pressure, and improve sleep.

Activities that increase cardiorespiratory fitness include brisk walking, jogging, bicycling, and swimming. Other sports, such as tennis, basketball, and racquetball, also increase cardiorespiratory fitness. Activities such as rowing or circuit training have the dual benefit of increasing both cardiorespiratory fitness and strength fitness.

WELL Building Standard Features: Fitness Concept						
Features	Project Type			Verification Documentation		
	Core and Shell	New and Existing Interiors	New and Existing Buildings	Letter of Assurance	Annotated Documents	On-Site Assessment
64, Interior Fitness Circulation						
Part 1: Stair Accessibility and Promotion	P	O	P			Visual Inspection
Part 2: Staircase Design	P	O	P			Visual Inspection
Part 3: Facilitative Aesthetics	P	O	P			Visual Inspection
65, Activity Incentive Programs						
Part 1: Activity Incentive Programs	–	P	P		Policy Document	
66, Structured Fitness Opportunities						
Part 1: Fitness Programs	–	O	O		Policy Document	
Part 2: Fitness Education	–	O	O		Policy Document	
67, Exterior Active Design						
Part 1: Pedestrian Amenities	O	O	O	Owner		Spot Check
Part 2: Pedestrian Promotion	O	O	O	Owner		Spot Check
Part 3: Neighborhood Connectivity	O	O	O		Annotated Map	
68, Physical Activity Spaces						
Part 1: Site Space Designation for Offices	O	O	O		Architectural Drawing	
Part 2: External Exercise Spaces	O	O	O		Annotated Map	
69, Active Transportation Support						
Part 1: Bicycle Storage and Support	O	O	O	Owner		Spot Check
Part 2: Post Commute and Workout Facilities	O	O	O	Architect		Spot Check
70, Fitness Equipment						
Part 1: Cardiorespiratory Exercise Equipment	O	O	O	Owner		Spot Check
Part 2: Muscle-Strengthening Exercise Equipment	O	O	O	Owner		Spot Check
71, Active Furnishings						
Part 1: Active Workstations	–	O	O	Owner		Spot Check
Part 2: Prevalent Standing Desks	–	O	O	Owner		Spot Check

Figure 9-1. The Fitness Concept in WELL includes eight features that promote strategies such as interior and exterior design, structured fitness opportunities, and the proper space and equipment for fitness activities to increase the level of fitness for building occupants.

TYPES OF FITNESS

Precor

Figure 9-2. The three types of fitness are strength and endurance, cardiorespiratory, and flexibility fitness.

Flexibility Fitness

Flexibility fitness addresses a body's ability to move. Increased flexibility fitness provides an increased range of motion for the musculoskeletal system. Increased flexibility reduces the chances of injury, decreases muscle soreness, increases balance and coordination, and may improve posture. The most basic way to increase flexibility is to stretch. Activities such as yoga, Pilates, and tai chi can also increase flexibility.

WELLNESS FACT

A Harvard University study found that average employee health care costs fell by $3.27 for every $1.00 spent on wellness programs.

FITNESS ADVANTAGES

Several advantages can be gained for both owners and employees when employees increase their fitness. Owners or employers that provide employee fitness programs may see lower healthcare and disability costs, increased employee productivity, reduced employee absences, improved morale, and improved employee recruitment and retention. Employees who use employer fitness programs may see decreased stress levels, lower out-of-pocket expenses for healthcare services, increased well-being and self-esteem, and improved job satisfaction.

FEATURE 64, INTERIOR FITNESS CIRCULATION

One of the easiest and most convenient methods of increasing fitness in a built environment, such as a commercial office building, is stair climbing. Since stairways are required building features in many commercial structures, WELL leverages this preexisting design necessity to encourage occupants to use stairs to increase their fitness and reduce sedentary behavior. *Sedentary behavior* is a manner of activity that involves sitting or lying down and is characterized by low levels of energy expenditure.

Feature 64, Interior Fitness Circulation, is a precondition for the Core and Shell and the New and Existing Buildings project types and an optimization for the New and Existing Interiors project type. The three parts of this feature are stair accessibility and promotion, staircase design, and facilitative aesthetics.

Part 1: Stair Accessibility and Promotion

Part 1: Stair Accessibility and Promotion of Feature 64 applies to projects with two to four floors. Part 1a requires that at least one common staircase be accessible to regular building occupants during all regular business hours. A building with only one tenant occupying the entire building can meet this requirement relatively easily. However, in a building with multiple tenants, there may be security concerns for the stairways and other shared common spaces. In this case, a building control plan should be in place during the WELL implementation process to ensure that the accessibility requirements can be met.

Part 1b requires the presence of wayfinding signage throughout the space. Point-of-decision prompts must be present to encourage stairway use. Wayfinding signage must be present at elevator banks (at least one sign per bank) to assist building occupants in locating a stairway. **See Figure 9-3.** The signage should meet the local accessibility guidelines.

Part 2: Staircase Design

Part 2: Staircase Design of Feature 64 applies to projects with two to four floors. At least one common staircase in the project must meet the three requirements of Part 2. **See Figure 9-4.** These requirements include the following:

- The staircase must be located within 7.5 m (25′) of the main entrance, main entry checkpoint, edge of the lobby, or main welcome area, which are high-traffic routes.
- The staircase must be clearly visible from those high-traffic routes or must be located visually before any elevators upon entering from the main entrance.
- The staircase must have a minimum width of 1.4 m (56″), or the width allowable by local code, between handrails.

STAIRCASE DESIGN

Figure 9-4. For Part 2: Staircase Design of Feature 64, Interior Fitness Circulation, a common staircase must meet location, visibility, and width requirements.

STAIR ACCESSIBILITY SIGNAGE

StepJockey Smart Signs

Figure 9-3. Signage at elevator banks should encourage stairway use and assist building occupants in locating a compliant stairway.

Part 3: Facilitative Aesthetics

Part 3: Facilitative Aesthetics of Feature 64 applies to projects with two to four floors. This part requires that both common stairways and paths of frequent travel include visual or auditory aesthetic elements to help make using stairways for fitness purposes more appealing. **See Figure 9-5.** To achieve Part 3, a project must incorporate at least two of the following elements: artwork, music, daylighting using windows or skylights at least 1 m^2 (10.8 ft^2) in total area, view windows to the outdoors or the building interior, light levels of at least 215 lux (20 fc) when stairs are being used, or biophilic elements.

FACILITATIVE AESTHETICS

Figure 9-5. Elements such as artwork, daylighting, and appropriate light levels make stair use more appealing.

FEATURE 65, ACTIVITY INCENTIVE PROGRAMS

Feature 65, Activity Incentive Programs, is a precondition for New and Existing Interiors and New and Existing Buildings project types. This precondition requires employers to develop a fitness incentive plan that covers all employees. Employers use the incentive plan to provide their employees with reimbursements or subsidies to promote various healthy activities. Employers should consult with their employees to determine which strategies work best.

Part 1: Activity Incentive Programs

Part 1: Activity Incentive Programs of Feature 65 requires the development and implementation of an activity incentive plan that includes at least two strategies. These strategies include the following:

- tax-exempt payroll deductions (or direct subsidies) relating to active transportation or mass transit use
- meaningful reimbursements or incentive payments (including nonmonetary) offered for every six-month period that an employee meets a 50-visit minimum to a gym or physical activity program
- a meaningful subsidy offered at least yearly towards participation or membership costs for fitness activities such as races, group fitness classes, sports teams, fitness centers, training centers, gyms, or studios
- a meaningful subsidy offered at least yearly towards the cost of an annual bicycle share membership
- no-cost or discounted physical activity opportunities or memberships, in which it can be demonstrated that 30% of occupants have utilized on a regular basis (at least weekly) over the last six months

However, the strategies should only be parts of the complete plan. A well-developed plan may include other incentives to keep employees engaged in healthy lifestyles. Other incentives that may be included in the plan are smoking cessation programs, wellness screenings, and wellness coaching. Some companies utilize wearable fitness trackers for voluntary company-wide wellness competitions.

FEATURE 66, STRUCTURED FITNESS OPPORTUNITIES

Not all employees may know how to start a fitness program, or they may have special considerations that may require outside assistance. Feature 66, Structured Fitness Opportunities, is an optimization for the New and Existing Interiors and New and Existing Buildings project types. The goal of this feature is to provide professional guidance and education to employees. **See Figure 9-6.** Guidance and education can be offered as a group class or as individual instruction.

Part 1: Fitness Programs

Part 1: Fitness Programs of Feature 66 requires that on-site fitness or training programs be offered from a qualified professional at least once per month. On-site fitness or training programs may include in-person education, group fitness classes, or online courses. The subjects of these programs may include general wellness and nutrition, behavioral changes, such as weight loss or smoking cessation, or targeted fitness goals, such as training for a 10 km race.

Part 2: Fitness Education

Part 2: Fitness Education of Feature 66 requires that targeted fitness training sessions be offered at least once every three months. These training sessions must be conducted by a qualified professional, such as a personal trainer or coach. The training sessions must also cover the following subjects:

- different modes of exercise, such as aerobic training and strength training, that increase endurance, strength, balance, or flexibility
- safe fitness techniques to help reduce the chance of injuries that could sideline employees' fitness progress
- comprehensive exercise regimens that address employees' specific goals, such as weight loss, stress reduction, or competitive training

BY THE NUMBERS

- *In the United States alone, fewer than 50% of elementary school students, 10% of adolescents, and 5% of adults obtain 30 minutes of daily physical activity.*

STRUCTURED FITNESS OPPORTUNITIES

Concept2, Inc.

Figure 9-6. Feature 66, Structured Fitness Opportunities, is an optimization requiring that professional guidance and education be provided to employees.

FEATURE 67, EXTERIOR ACTIVE DESIGN

Feature 67, Exterior Active Design, is an optimization for all three project types. The goal of this feature is to provide building occupants increased opportunities for outdoor activity. The first two parts of this feature apply to projects where the building footprint takes up less than 75% of the total lot size. If the building footprint is more than 75%, then only Part 3 applies.

Part 1: Pedestrian Amenities

A *pedestrian amenity* is a design feature that provides functional services for building occupants and visitors while at the same time making a space more comfortable and engaging. For Part 1: Pedestrian Amenities of Feature 67, at least one of three options must be provided in high-traffic areas that include building entrances, public transportation stops, walking paths, and plazas. These options include a bench, a cluster of movable tables and chairs, or a drinking fountain or water refilling station. **See Figure 9-7.**

PEDESTRIAN AMENITIES

Figure 9-7. Options for Part 1: Pedestrian Amenities of Feature 67, Exterior Active Design, include a bench, a cluster of movable tables and chairs, or a drinking fountain or water refilling station.

Part 2: Pedestrian Promotion

Part 2: Pedestrian Promotion of Feature 67 encourages outdoor activity by requiring design features that visually stimulate pedestrians and other users of the outdoor space. At least two features must be included outdoors on the site. These design features include a water fountain or feature, a plaza or open-air courtyard, a garden or other landscaped elements, or public art.

Part 3: Neighborhood Connectivity

Part 3: Neighborhood Connectivity of Feature 67 encourages building occupants to engage the neighborhoods around their project sites. A building must meet at least one of two requirements. Both requirements for Part 3 involve metrics outside the WELL standard.

The first option, Part 3a, requires that a building's address achieve a Walk Score® of 70 or higher. *Walk Score®* is a measurement on a scale of 0 to 100 that takes into account a building inhabitant's physical output. **See Figure 9-8.** Walkability is determined by the amount of amenities within a given distance of a building's address. The closer that amenities are to the site, the higher the score achieved by the building.

Neighborhood Connectivity—Walk Score

Walk Score®	Description
90–100	**Walker's Paradise** Daily errands do not require a car
70–89	**Very Walkable** Most errands can be accomplished on foot
50–69	**Somewhat Walkable** Some errands can be accomplished on foot
25–49	**Car-Dependent** Most errands require a car
0–24	**Car-Dependent** Almost all errands require a car

Figure 9-8. Walk Score® is a measurement that takes into account a building inhabitant's physical output, measured on a scale of 0 to 100.

WELLNESS FACT

Because LEED and WELL share a common philosophy of sustainability and supporting human and environmental health, the IWBI, the USGBC, and GBCI have developed a crosswalk document (available at wellcertified.com) that helps project teams achieve both of these certifications.

The second option, Part 3b, requires that at least four existing and publicly available diverse uses are located within an 800 m (0.5 mile) walking distance of the building's main entrance. A *diverse use* is a distinct business or organization that provides goods or services intended to meet daily needs and is publicly available. **See Figure 9-9.** Acceptable examples of diverse uses are found in USGBC's *LEED Reference Guide for Building Design and Construction* (LEED BD+C), Appendix 1: Use Types and Categories. Some examples of diverse uses include grocery stores, pharmacies, banks, gyms, childcare facilities, places of worship, and public parks.

FEATURE 68, PHYSICAL ACTIVITY SPACES

Building occupants who seek fitness opportunities should be provided spaces for both indoor and outdoor physical activities. A combination of designated indoor and outdoor exercise spaces provides more choices for physical activity and allows for continued activity year-round, regardless of weather conditions. Feature 68, Physical Activity Spaces, is an optimization for all three project types.

BY THE NUMBERS

- *An average adult obtains only 6 minutes to 10 minutes of moderate to vigorous intensity physical activity per day.*
- *Just 2.5 hours of moderate-intensity physical activity per week can reduce overall mortality risk by nearly 20%.*

DIVERSE USES

Figure 9-9. The more diverse uses there are within walking distance of a project, the greater incentive building occupants have to utilize physical activity to meet their daily needs.

Part 1: Site Space Designation for Offices

A dedicated indoor exercise space on a site allows for physical activity all year long. A designated space with complimentary access will help building occupants maintain a consistent exercise program due to its convenience. Part 1: Site Space Designation for Offices of Feature 68 applies to spaces with more than 10 regular occupants.

The dedicated exercise space must have a minimum area of 18.6 m^2 (200 ft^2) plus 0.1 m^2 (1 ft^2) per regular building occupant. **See Figure 9-10.** For example, a designated fitness space for 100 regular building occupants would have to be at least 28.6 m^2 (300 ft^2) $[18.6 + (100 \times 0.1) = 28.6]$. The maximum size for a dedicated exercise space is 370 m^2 (4000 ft^2).

Part 2: External Exercise Spaces

Building occupants who are given choices for exercise spaces are likely to take advantage of fitness opportunities. Part 2: External Exercise Spaces of Feature 68 provides several options for external exercise spaces. Part 2 requires that at least one external exercise space be located within 0.8 km (0.5 mile) walking distance of the building's main entrance and have complimentary access. The external exercise spaces that can meet these requirements are a green space or park with playground features; a workout station or fitness zone; a trail network; an accessible body of water or public swimming pool; a gym, fitness, or training center; or a recreational field.

FEATURE 69, ACTIVE TRANSPORTATION SUPPORT

Active transportation is a form of commuting by way of a physical activity such as biking or walking. Active transportation is associated with multiple health benefits and can reduce greenhouse gas emissions by reducing reliance on automobiles for commuting. If a project's location is conducive to active transportation, the project team may provide support facilities for the building occupants who choose active commuting. Feature 69, Active Transportation Support, is an optimization for all three project types.

Part 1: Bicycle Storage and Support

Part 1: Bicycle Storage and Support of Feature 69 requires that bicycle maintenance tools and both short- and long-term bicycle storage be available for occupant use. Both the maintenance tools and the storage spaces must be located on-site or within 200 m (650′) of the building's main entrance. The maintenance tools should include basic items such as tire pumps, patch kits for tube repair, and hex keys for mechanical adjustments.

Long-term bicycle storage space that is separate and secure must be sized to accommodate 5% of the building's regular occupants. The short-term bicycle storage space must accommodate 2.5% of the building's peak visitors. For example, if the building has 120 regular occupants and 50 average peak visitors, there must be at least six long-term bicycle storage spaces ($120 \times 5\% = 6$) and at least two short-term bicycle storage spaces ($50 \times 2.5\% = 1.25$, rounded up to 2).

Part 2: Post Commute and Workout Facilities

When building occupants are provided shower facilities and lockers for use on-site, they are more likely to consider active transportation or participation in other fitness activities. **See Figure 9-11.** Similar to the requirements for bicycle storage spaces, Part 2: Post Commute and Workout Facilities of Feature 69 requires that shower facilities and lockers be located on-site or within 200 m (650′) of the building's main entrance.

One shower with a changing facility is required for the first 100 regular building occupants and one additional shower for every 150 regular occupants after that. One locker is required for every five building occupants. For example, for a project with 125 regular occupants, at least 2 showers with changing facilities and 25 lockers

(125 ÷ 5 = 25) are required. The number of lockers may differ from the prescribed ratio if there is evidence that the number of lockers provided exceed demand by 20%.

FEATURE 70, FITNESS EQUIPMENT

Feature 70, Fitness Equipment, is an optimization for all three project types that addresses the availability of fitness equipment. The goal of this feature is to provide the building's occupants with complimentary access to the proper exercise equipment for meeting their individual fitness goals. This feature contains two parts that address cardiorespiratory and muscle-strengthening exercise equipment, respectively.

Both parts of this feature require a combination of the listed equipment so that at least 1% of regular building occupants can use the equipment. For example, a project with 150 building occupants would require two pieces of equipment from Part 1 (150 × 1% = 1.5, rounded up to 2) and two pieces of equipment from Part 2 (150 × 1% = 1.5, rounded up to 2). There can be no charge for equipment use, and instructions for safe use must be included.

POST COMMUTE AND WORKOUT FACILITIES

Figure 9-11. Part 2: Post Commute and Workout Facilities of Feature 69, Active Transportation Support, requires that shower facilities and lockers be available for occupant use on-site.

DEDICATED EXERCISE SPACE

Figure 9-10. A dedicated indoor exercise space on a site allows for year-round physical activity and must be a minimum of 18.6 m^2 (200 ft^2) plus 0.1 m^2 (1 ft^2) per regular building occupant.

Part 1: Cardiorespiratory Exercise Equipment

Part 1: Cardiorespiratory Exercise Equipment of Feature 70 requires that exercise equipment designed to increase cardiorespiratory fitness be provided for regular building occupants. When this equipment is provided for indoor use, building occupants are able to increase their fitness year-round. The equipment can be used for activities such as walking, running, biking, and rowing. **See Figure 9-12.** If performed correctly, these activities contribute to increased cardiorespiratory fitness and may also contribute to increased muscle strength as a secondary benefit.

The following four types of cardiorespiratory equipment may be used in combination to satisfy the requirements of Part 1:

- **Treadmills.** A treadmill is used for walking or running by a user. A treadmill often includes the ability to increase the speed and incline of its running belt to meet the user's needs.
- **Elliptical machines.** An elliptical machine is used to simulate walking, running, or stair climbing but reduces the amount of pressure and stress on the lower body muscles and joints.
- **Rowing machines.** A rowing machine, sometimes referred to as an ergometer, is used to simulate rowing a boat on

CARDIORESPIRATORY EXERCISE EQUIPMENT

Precor

Concept2, Inc.

Figure 9-12. Cardiorespiratory exercise equipment, such as treadmills, elliptical machines, rowing machines, and stationary exercise bicycles, provide increased cardiorespiratory fitness when used correctly.

water. Proper technique is important on a rowing machine in order to reduce the chance of injury.

- **Stationary exercise bicycles.** A stationary exercise bicycle is used to simulate biking. There are two types of stationary exercise bicycles: upright and recumbent. An upright exercise bicycle simulates a traditional road bicycle. A recumbent exercise bicycle allows the user to sit lower and in a semireclined position.

BY THE NUMBERS

- *Over 60% of all people worldwide do not get the recommended daily 30-minute minimum of moderate-intensity physical activity.*
- *Physical inactivity is an independent risk factor for numerous chronic diseases and is estimated to be responsible for 30% of ischemic heart disease (heart problems caused by narrowed arteries), 27% of type 2 diabetes, and 21% to 25% of breast and colon cancer cases.*

Part 2: Muscle-Strengthening Exercise Equipment

Part 2: Muscle-Strengthening Exercise Equipment of Feature 70 requires that exercise equipment designed for increasing muscle strength be provided to regular building occupants. **See Figure 9-13.** The following four types of muscle-strengthening exercise equipment may be used in combination to satisfy the requirements of Part 2:

- **Multistation equipment.** Multistation equipment uses a combination of weights, pulleys, and cables to allow a user to perform multiple exercises on a single machine. Although multistation equipment replaces several pieces of exercise equipment, it may have a considerable footprint.
- **Bench-press with a self-spotting rack.** A bench press machine is a bench that is used for developing muscles groups in the chest, shoulders, and arms. A self-spotting rack allows the use of the equipment without the need for an additional person to act as a spotter.
- **Full squat-rack.** A full squat-rack is used to safely perform squats and other lower body and back exercises.
- **Pull-up bar.** A pull-up bar, available either as a self-supported type or a wall-mounted type, is used for pull-ups and other exercises.

FEATURE 71, ACTIVE FURNISHINGS

Many jobs require a person to staff a workstation. Staffing a workstation usually involves sitting in a chair for the entire workday. This sedentary behavior can cut years from a person's life expectancy. Sitting slows the circulation of blood through the body and can increase the risk of cardiovascular disease, cancer, and obesity.

Feature 71, Active Furnishings, is an optimization for the New and Existing Interiors and New and Existing Buildings project types. The goal of this feature is to reduce the amount of prolonged sitting and increase the amount of physical activity throughout the day. The first part of this optimization feature requires that specialized workstations be provided to increase employees' physical activity while working. The second part requires that workstations allowing for height variation during their use be provided for employees.

Adjustable-height workstations accommodate the needs of a variety of users.

MUSCLE-STRENGTHENING EXERCISE EQUIPMENT

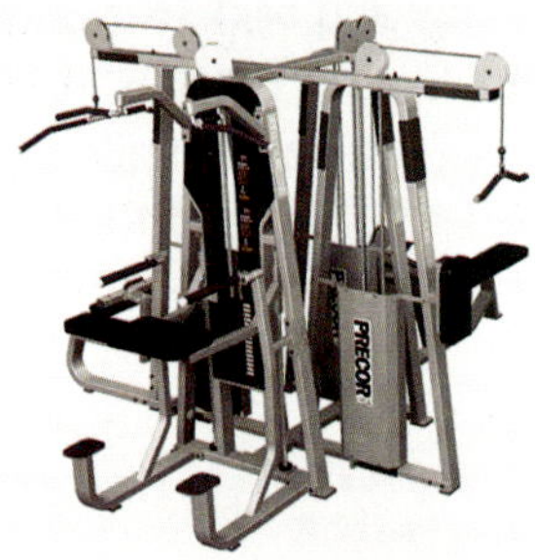

Multistation Equipment

Bench Press with Self-Spotting Rack

Full Squat-Rack

Pull-Up Bar

Precor

Figure 9-13. Part 2: Muscle-Strengthening Exercise Equipment of Feature 70, Fitness Equipment, requires that exercise equipment designed to increase muscle strength be provided to building occupants.

BY THE NUMBERS

- *Sitting burns 50 fewer calories per hour than standing, and sitting for more than 3 hours per day is associated with a 2-year lower life expectancy.*

Part 1: Active Workstations

Part 1: Active Workstations of Feature 71 requires that treadmill desks, bicycle desks, or portable desk pedal or stepper machines be provided for at least 3% of the employees (with a minimum of one). **See Figure 9-14.** For example, a project that has 125 employees must have at least four machines (125 × 3% = 3.75, rounded up to 4) in any combination.

ACTIVE WORKSTATIONS

Steelcase

Figure 9-14. Part 1: Active Workstations of Feature 71, Active Furnishings, requires that treadmill desks, bicycle desks, or portable under-desk peddlers or stepper machines be provided for at least 3% of the employees (with a minimum of one).

Part 2: Prevalent Standing Desks

Part 2: Prevalent Standing Desks of Feature 71 requires that 60% of the workstations, not employees, have either an adjustable-height standing desk, a standard desk with a desktop height adjustment stand, or pairs of fixed-height desks of standing and seated heights. **See Figure 9-15.** For example, if a project has 85 workstations, then 51 workstations (85 × 60% = 51) are required to meet Part 2.

BY THE NUMBERS

- *Physical inactivity is the fourth leading risk factor for mortality, accounting for 6% to 9% of deaths worldwide, or 3 million to 5 million mortalities every year.*
- *Lack of physical activity can increase the odds of having a stroke by 20% to 30% and shave off three to five years of life.*

PREVALENT STANDING DESKS

Steelcase

Figure 9-15. Part 2: Prevalent Standing Desks of Feature 71, Active Furnishings, requires that 60% of the workstations have either an adjustable-height standing desk, a standard desk with a desktop height adjustment stand, or pairs of fixed-height desks of standing and seated heights.

KEY TERMS AND DEFINITIONS

active transportation: A form of commuting by way of a physical activity such as biking or walking.

cardiorespiratory fitness: A component of physical fitness that involves the ability of the body, specifically the heart and lungs, to transport, absorb, and use oxygen during sustained physical activity. Also referred to as aerobic fitness.

diverse use: A distinct business or organization that provides goods or services intended to meet daily needs and is publicly available.

fitness: The ability to carry out daily tasks with vigor and alertness, without undue fatigue, and with ample energy to enjoy leisure-time pursuits and respond to emergencies.

pedestrian amenity: A design feature that provides functional services for building occupants and visitors while at the same time making a space more comfortable and engaging.

sedentary behavior: A manner of activity that involves sitting or lying down and is characterized by low levels of energy expenditure.

Walk Score®: A measurement on a scale of 0 to 100 that takes into account a building inhabitant's physical output.

CHAPTER REVIEW

Completion

________________ **1.** Strength and endurance fitness exercises are also known as ___ training.

________________ **2.** ___ is a manner of activity that involves sitting or lying down and is characterized by low levels of energy expenditure.

________________ **3.** Feature 64, Interior Fitness Circulation, applies to projects with ___ floors.

________________ **4.** In order to utilize the first two parts of Feature 67, Exterior Active Design, the building footprint must take up less than ___% of the total lot size.

________________ **5.** A(n) ___ is a design feature that provides functional services for building occupants and visitors while at the same time making a space more comfortable and engaging.

________________ **6.** In order to qualify for Part 3a of Feature 67, Exterior Active Design, a building's address must have a Walk Score® of ___ or higher.

________________ **7.** In order to meet Part 3b: Neighborhood Connectivity of Feature 67, Exterior Active Design, a project must have at least ___ existing and publicly available diverse uses located within an 800 m (0.5 mile) walking distance of the building's main entrance.

________________ **8.** A(n) ___ is a distinct business or organization that provides goods or services intended to meet daily needs and is publicly available.

________________ **9.** The maximum size for a dedicated exercise space for Part 1 of Feature 68, Physical Activity Spaces, is ___.

CHAPTER REVIEW *(continued)*

______________________ **10.** ___ is a form of commuting by way of a physical activity such as biking or walking.

______________________ **11.** There must be one locker for every ___ regular building occupants for Part 2: Post Commute and Workout Facilities of Feature 69, Active Transportation Support.

______________________ **12.** Treadmills and elliptical machines are examples of ___ exercise equipment.

______________________ **13.** A full squat-rack and a pull-up bar are examples of ___ exercise equipment.

______________________ **14.** Part 1: Active Workstations of Feature 71, Active Furnishings, requires that treadmill desks, bicycle desks, or portable desk pedal or stepper machines be provided for ___% or more of employees.

______________________ **15.** For Part 2: Prevalent Standing Desks of Feature 71, Active Furnishings, ___% of workstations must feature standing desks, height-adjustable desktop stands, or pairs of fixed-height desks of standing and seated heights.

Short Answer

1. What are the three requirements that at least one common staircase in a project must meet per Part 2: Staircase Design of Feature 64, Interior Fitness Circulation?

2. List the five elements that are options for fulfilling Part 3: Facilitative Aesthetics of Feature 64, Interior Fitness Circulation.

3. Feature 65, Activity Incentive Programs, is a precondition for which project types?

4. List the two parts of Feature 66, Structured Fitness Opportunities.

5. List the three amenities that qualify for Part 1: Pedestrian Amenities of Feature 67, Exterior Active Design.

6. List the four elements that qualify for Part 2: Pedestrian Promotion of Feature 67, Exterior Active Design.

CHAPTER REVIEW *(continued)*

7. List the spaces that can be used for Part 2: External Exercise Spaces of Feature 68, Physical Activity Spaces, if they are within 0.8 km (0.5 mile) walking distance to the building.

8. What must be provided on-site or within 200 m (650′) of a building's main entrance for Part 1: Bicycle Storage and Support of Feature 69, Active Transportation Support?

9. What two types of fitness equipment are required by Feature 70, Fitness Equipment?

10. Part 1: Active Workstations of Feature 71, Active Furnishings, requires active workstations for what percentage of employees?

WELL AP EXAM PRACTICE QUESTIONS

1. Which of the following qualifies for Part 3: Facilitative Aesthetics of Feature 64, Interior Fitness Circulation?

A. Artwork, music, and views to the outdoors
B. A water feature and light levels of at least 215 lux (20 fc)
C. Daylighting using windows 0.75 m^2 (8 ft^2) in size, music, and artwork
D. Music, artwork, and light levels of at least 200 lux (18.6 fc)

2. What does Feature 69, Active Transportation Support, require on-site or within 200 m (650′) of a building's main entrance?

A. Basic bicycle maintenance tools, short-term bicycle storage, a changing facility, and lockers
B. Short- and long-term bicycle storage, a water bottle refilling station, and lockers
C. Basic bicycle maintenance tools, short- and long-term bicycle storage, and lockers
D. Showers and changing facilities, a dedicated workout space, and short- and long-term bicycle storage

3. What part and feature in the Fitness Concept require an architectural drawing to be submitted to demonstrate that the part's requirements have been satisfied?

A. Part 3: Facilitative Aesthetics of Feature 64, Interior Fitness Circulation
B. Part 1: Activity Incentive Programs of Feature 66, Activity Incentive Programs
C. Part 1: Site Space Designation for Offices of Feature 68, Physical Activity Spaces
D. Part 1: Active Workstations of Feature 71, Active Furnishings

4. What combination of exercise equipment meets the minimum requirements of Feature 70, Fitness Equipment, for a building with 45 regular building occupants?

A. One treadmill, two elliptical machines, two rowing machines, and two pull-up bars
B. Two squat-racks, two pull-up bars, three treadmills, and two stationary bikes
C. Two treadmills, three elliptical machines, one rowing machine, one multistation equipment set, and one squat-rack
D. Two stationary bikes, three treadmills, one squat-rack, one bench-press rack, one multistation equipment set, and two pull-up bars

5. Bicycle storage and showers with changing rooms are two major space considerations for which feature?

A. Feature 68, Physical Activity Spaces
B. Feature 69, Active Transportation Support
C. Feature 70, Fitness Equipment
D. Feature 71, Active Furnishings

CHAPTER TEN

COMFORT

Although comfort can be a subjective issue, the built environment plays a major role in the perception of each person's level of comfort. The perceived level of comfort in the built environment affects the productivity and well-being of building occupants. Building occupants are able to perform work to the best of their abilities when the indoor environment, such as an office building, is comfortable and free of distractions. The goal of the Comfort Concept in the WELL Building Standard is to design the acoustics, ergonomics, and thermal environmental conditions of a building to prevent stress and eliminate unwanted distractions.

OBJECTIVES

- Identify the features in the Comfort Concept as preconditions or optimizations.
- Explain how acoustics, ergonomics, olfaction, and thermal comfort impact the level of comfort in the built environment and the well-being of building occupants.
- Explain how ergonomic hazards can lead to musculoskeletal disorders.
- Identify the six factors of thermal comfort per ASHRAE Standard 55.
- Identify the law that prohibits discrimination against persons with disabilities and ensures that an equitable built environment is designed.
- Describe the ergonomic design strategies that help prevent musculoskeletal disorders and improve building occupant comfort.
- Describe the strategies for minimizing sound distractions and achieving better acoustic comfort.
- Explain the different metrics used to measure the qualities of sound and sound transmission.
- Describe the strategies for controlling the thermal environment of a building.
- Explain the concept of free address and its role in thermal comfort.

KEY TERMS

- acoustics
- Americans with Disabilities Act (ADA)
- decibel (dB)
- electric radiant system
- ergonomics
- free address
- hydronic heating and/or cooling system
- musculoskeletal disorder
- noise criteria (NC)
- noise insulation class (NIC)
- noise reduction coefficient (NRC)
- olfaction
- olfactory comfort
- psychrometric chart
- psychrometrics
- reverberation time
- sound
- sound masking
- sound pressure level
- sound transmission class (STC)
- thermal comfort

Learner Resources

ATPeResources.com/QuickLinks
Access Code: 935534

COMFORT CONCEPT

When building occupants are uncomfortable for whatever reason, they cannot perform at their peak efficiency. Not only does the lack of comfort cause both psychological and physiological disruptions, the smallest distractions can be compounded by feelings of discomfort. WELL aims to maintain a healthy and productive work environment for all of a building's occupants and visitors by addressing areas of comfort such as acoustics, ergonomics, olfaction, and thermal comfort.

The Comfort Concept in WELL includes 12 features that promote strategies such as accessibility, noise reduction, thermal comfort and control, and olfactory comfort. Core and Shell projects must meet three preconditions, but two optimizations can be pursued for higher certification. New and Existing Interiors must meet four preconditions, but eight optimizations can be pursued for higher certification. New and Existing Buildings must meet five preconditions, but seven optimizations can be pursued for higher certification. **See Figure 10-1.**

AREAS OF COMFORT

Several areas of comfort are addressed by the features of the Comfort Concept. These areas include acoustics, ergonomics, olfaction, and thermal comfort. Each of these areas plays an important role in maintaining a satisfactory level of comfort in a building.

Acoustic Comfort

Acoustics is the study of sound and the properties of a space or building that determine how sound is transmitted or reflected within it or through it. *Sound* is energy composed of pressure waves, or vibrations, that travel through the air or another material and produce an audible signal when they reach a person's ear. These pressure waves can be measured in hertz (Hz), which is the unit of measure used for the frequency that a wave cycles, or repeats, its wavelength. One hertz is equal to one cycle per second (1 Hz = 1 cycle/second). The typical human ear can hear sound in frequencies from 20 Hz to 20,000 Hz. **See Figure 10-2.**

In some instances, a sound can be detrimental to the person hearing it. Detrimental sound is referred to as noise. Noise can cause physical and psychological stress, which leads to cardiovascular disease, decreased productivity and cognitive performance, and difficulty in communication and concentration. Noise experienced during the day may even disturb sleep cycles at night. Excessively loud noise, especially over extended periods, can cause hearing damage and loss.

Ergonomic Comfort

Ergonomics is the science of adapting objects, spaces, and processes, such as workstations and workflow, to accommodate people's capabilities in a safe and efficient manner. Examples of ergonomic hazards in office buildings include limited office chair capabilities, uncomfortable desk and table heights, and poor body positioning. Repetitive exposure to ergonomic hazards can lead to increased stress, lower productivity, and musculoskeletal disorders.

Musculoskeletal Disorders. A *musculoskeletal disorder* is an injury or condition that affects the body's movements or its muscles, bones, tendons, nerves, or ligaments. Common musculoskeletal disorders that can occur in an office building include carpal tunnel syndrome in the hands and arms, tendonitis in the hands and wrists, and pain in the back, hips, or neck. Any of these disorders can affect the well-being of an occupant, reduce productivity, and increase healthcare costs. Treatments for musculoskeletal disorders include the restriction of movement, the application of heat or cold treatments, physical therapy, or medication or surgery.

BY THE NUMBERS

- *In 2013, nearly 380,600 days of work were missed in the United States (one-third of all days away from work) due to musculoskeletal disorders.*

WELL Building Standard Features: Comfort Concept						
Features	Project Type			Verification Documentation		
	Core and Shell	New and Existing Interiors	New and Existing Buildings	Letter of Assurance	Annotated Documents	On-Site Assessment
72, Accessible Design						
Part 1: Accessibility and Usability	P	P	P	Architect		
73, Ergonomics: Visual and Physical						
Part 1: Visual Ergonomics	–	P	P	Owner		Spot Check
Part 2: Desk Height Flexibility	–	P	P	Owner		Spot Check
Part 3: Seat Flexibility	–	P	P	Owner		Spot Check
74, Exterior Noise Intrusion						
Part 1: Sound Pressure Level	P	O	P			Performance Test
75, Internally Generated Noise						
Part 1: Acoustic Planning	–	P	P		Professional Narrative	
Part 2: Mechanical Equipment Sound Levels	O	P	P			Performance Test
76, Thermal Comfort						
Part 1: Ventilated Thermal Environment	P	P	P	MEP		Spot Measurement
Part 2: Natural Thermal Adaptation	P	P	P	MEP		Spot Measurement
77, Olfactory Comfort						
Part 1: Source Separation	–	O	O		Architectural Drawing	
78, Reverberation Time						
Part 1: Reverberation Time	–	O	O			Performance Test
79, Sound Masking						
Part 1: Sound Masking Use	–	O	O	MEP		
Part 2: Sound Masking Limits	–	O	O			Performance Test
80, Sound Reducing Surfaces						
Part 1: Ceilings	–	O	O	Architect		
Part 2: Vertical Surfaces	–	O	O	Architect		
81, Sound Barriers						
Part 1: Wall Construction Specifications	–	O	O	Architect		
Part 2: Doorway Specifications	–	O	O	Architect & Contractor		
Part 3: Wall Construction Methodology	–	O	O	Contractor		
82, Individual Thermal Control						
Part 1: Free Address	–	O	O		Policy Document	
Part 2: Personal Thermal Comfort Devices	–	O	O	Owner		Spot Check
83, Radiant Thermal Comfort						
Part 1: Lobbies and Other Common Spaces	O	–	O	MEP		
Part 2: Offices and Other Regularly Occupied Spaces	–	O	O	MEP		

Figure 10-1. The Comfort Concept in WELL includes 12 features that promote strategies such as accessibility, noise reduction, thermal comfort and control, and olfactory comfort.

HUMAN HEARING

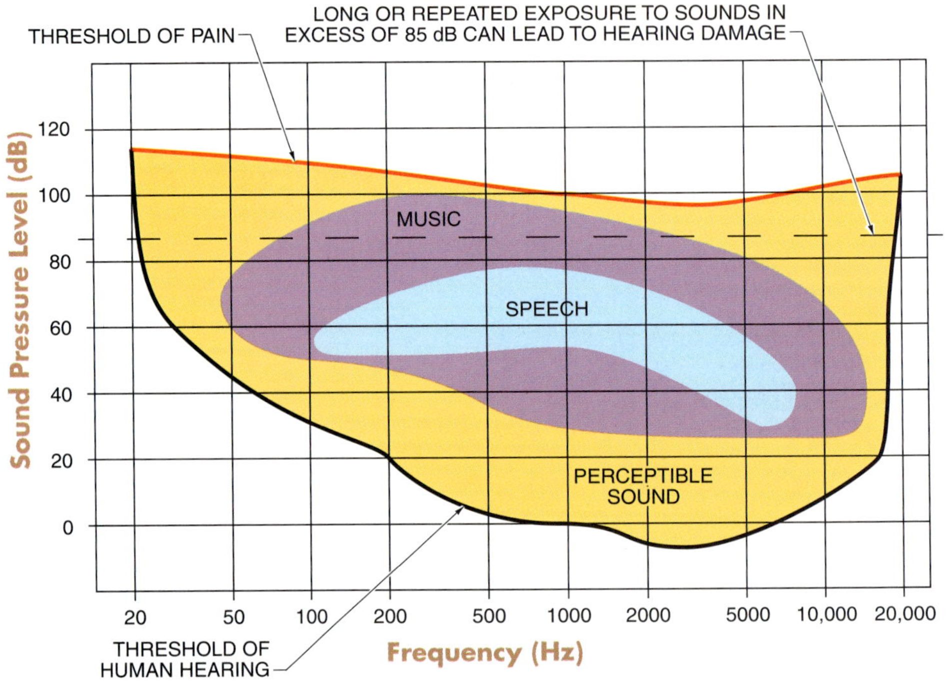

Figure 10-2. The typical human ear can hear sound in frequencies from 20 Hz to 20,000 Hz.

Olfactory Comfort

Olfaction is the sense of smell. *Olfactory comfort* is a person's perception of air quality within an environment based on their sense of smell. Unpleasant or overwhelming smells can trigger headaches, migraines, dizziness, allergic reactions, skin irritation, and asthma attacks. At the very least, these smells can cause mental distractions and increase stress. While it may not be possible to eliminate all of the unpleasant smells in a building, the spread of smells to other spaces of the building can be limited.

Thermal Comfort

Thermal comfort is the condition of satisfaction with the thermal environment. The standard for thermal comfort used in WELL is ASHRAE Standard 55-2013, *Thermal Environmental Conditions for Human Occupancy*. ASHRAE Standard 55-2013 specifies the following six factors for thermal comfort:

- **Metabolic rate.** Metabolic rate is the rate of transformation of chemical energy into heat and mechanical work by the metabolic activities of a person, expressed in units of met.
- **Clothing insulation.** Clothing insulation is the resistance to sensible heat transfer provided by a clothing ensemble, expressed in units of clo.
- **Air temperature.** Air temperature is the temperature of air surrounding a representative occupant taken with a dry bulb thermometer, measured in °C and °F.
- **Mean radiant temperature.** Mean radiant temperature is the average of the surface temperature of the surroundings with which the body can exchange heat by radiant transfer, measured in °C and °F.
- **Air speed.** Air speed is the rate of air movement at a point, without regard to direction, measured in m/s (fpm).

- **Humidity.** Humidity is the average moisture content of the surrounding air, measured in percent of relative humidity (% rh).

Building occupants can control the factors of metabolic rate and clothing insulation. The other four factors are dependent on a building's environment. In addition to these six factors, which are quantifiable metrics, occupant thermal comfort can be influenced by each individual occupant's expectations. These expectations can be measured using occupant surveys during the predesign phase all the way through the post-occupancy phase.

Psychrometric Charts. *Psychrometrics* is the mathematical and graphical study of the properties of air. A *psychrometric chart* is a graphical representation of the properties of air at various conditions. A psychrometric chart is used to determine the comfort zone for a given space using the six factors for thermal comfort. Web-based thermal comfort tools can be used to perform thermal comfort calculations and create psychrometric charts that comply with the requirements of ASHRAE Standard 55-2013. **See Figure 10-3.**

BY THE NUMBERS

- *Low back pain, one of the most common musculoskeletal disorders, affects about 31 million Americans at any given time.*

PSYCHROMETRIC CHARTS

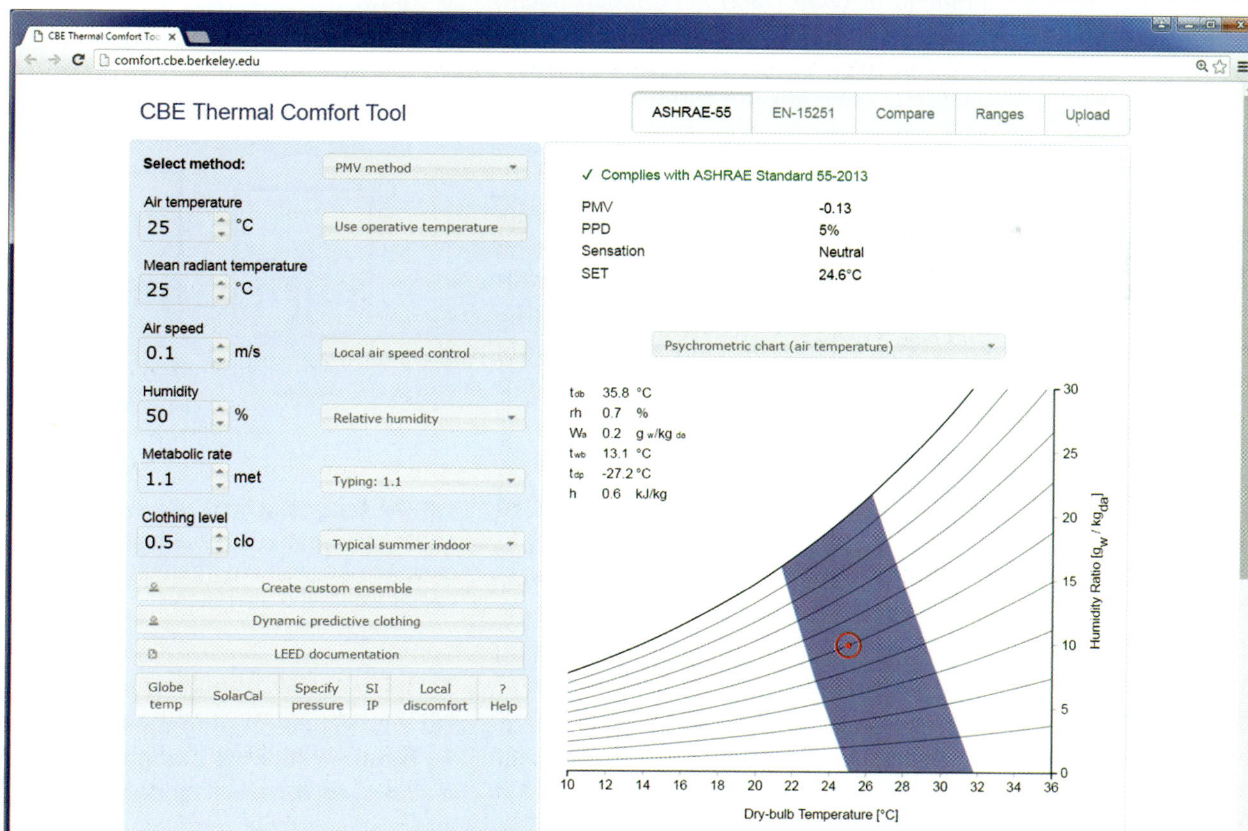

Center for the Built Environment

Figure 10-3. The Center for the Built Environment (CBE) Thermal Comfort Tool is a tool for thermal comfort calculations that displays results visually on a psychrometric chart.

FEATURE 72, ACCESSIBLE DESIGN

Feature 72, Accessible Design, is a precondition for all three project types that requires projects to provide an equitable built environment through compliance with one of two listed standards: the current ADA Standards for Accessible Design or ISO 21542:2011, Building Construction—Accessibility and Usability of the Built Environment. Comparable local codes or standards may also be used.

The *Americans with Disabilities Act (ADA)* is an enacted law that prohibits discrimination against persons with disabilities and ensures equal opportunity in employment, state and local government services, public accommodations, commercial facilities, and transportation. The *International Organization for Standardization (ISO)* is an independent, nongovernmental international organization that develops consensus-based, market-relevant standards for worldwide use.

Part 1: Accessibility and Usability

Part 1: Accessibility and Usability of Feature 72 requires that all projects comply with the current ADA Standards for Accessible Design or ISO 21542:2011, Building Construction—Accessibility and Usability of the Built Environment. The ADA Standards for Accessible Design is a document that sets the minimum accessibility requirements for new construction and alterations. **See Figure 10-4.** ISO 21542:2011 is an internationally recognized design standard that defines how built environments should be designed, constructed, and managed to meet the accessibility and usability needs of the majority of people.

FEATURE 73, ERGONOMICS: VISUAL AND PHYSICAL

Feature 73, Ergonomics: Visual and Physical, is a precondition for the New and Existing Interiors and New and Existing Buildings project types. When building occupants confine their body position to the static (nonmoving) components of their workstations on a daily basis, the repetitive discomfort can result in musculoskeletal disorders and lower productivity. The goal of this feature is to allow building occupants to adjust computer monitors, desk heights, and workstation seating to meet their personal needs and prevent injuries.

ADA ACCESSIBLE DESIGN STANDARDS

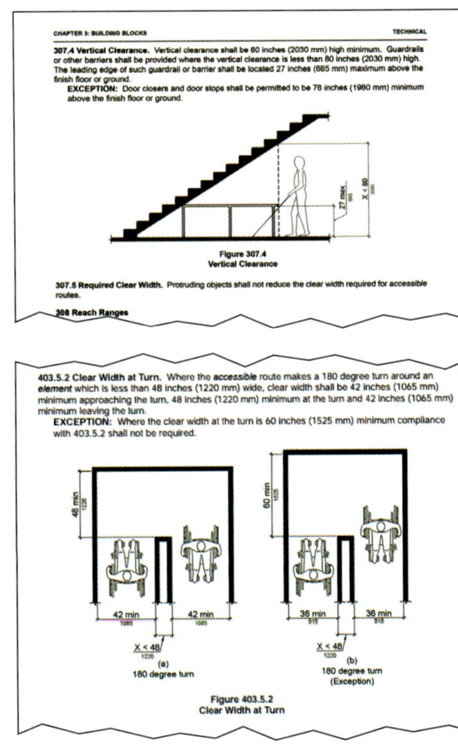
CHAPTER 3: BUILDING BLOCKS TECHNICAL

307.4 Vertical Clearance. Vertical clearance shall be 80 inches (2030 mm) high minimum. Guardrails or other barriers shall be provided where the vertical clearance is less than 80 inches (2030 mm) high. The leading edge of such guardrail or barrier shall be located 27 inches (685 mm) maximum above the finish floor or ground.

EXCEPTION: Door closers and door stops shall be permitted to be 78 inches (1980 mm) minimum above the finish floor or ground.

Figure 307.4
Vertical Clearance

307.5 Required Clear Width. Protruding objects shall not reduce the clear width required for *accessible* routes.

308 Reach Ranges

403.5.2 Clear Width at Turn. Where the *accessible* route makes a 180 degree turn around an *element* which is less than 48 inches (1220 mm) wide, clear width shall be 42 inches (1065 mm) minimum approaching the turn, 48 inches (1220 mm) minimum at the turn and 42 inches (1065 mm) minimum leaving the turn.

EXCEPTION: Where the clear width at the turn is 60 inches (1525 mm) minimum compliance with 403.5.2 shall not be required.

Figure 403.5.2
Clear Width at Turn

Figure 10-4. The ADA Standards for Accessible Design is a document that sets the minimum accessibility requirements for new construction and alterations.

Part 1: Visual Ergonomics

Part 1: Visual Ergonomics of Feature 73 requires that all computer screens, including laptops, be adjustable for height and distance from the user. This can be accomplished by having computer screens or laptops on adjustable risers on top of the desktop or by using adjustable articulated monitor supports. **See Figure 10-5.**

VISUAL ERGONOMICS

Steelcase

Figure 10-5. Adjustable articulated monitor supports can help satisfy Part 1: Visual Ergonomics of Feature 73, Ergonomics: Visual and Physical.

Part 2: Desk Height Flexibility

Part 2: Desk Height Flexibility of Feature 73 requires that at least 30% of workstations have the ability to alternate between sitting and standing positions using adjustable-height sit-stand desks, desktop height adjustment stands, or pairs of fixed-height desks of standing and seated heights. **See Figure 10-6.**

Part 3: Seat Flexibility

Part 3: Seat Flexibility of Feature 73 requires that the workstation chair height and seat depth adjustability be compliant with the HFES 100 standard or BIFMA G1 guidelines. **See Figure 10-7.**

HFES 100 is published by the Human Factors and Ergonomics Society (HFES), which is an interdisciplinary nonprofit professional organization covering the fields of human factors and ergonomics. HFES 100, *Human Factors Engineering of Computer Workstations,* is an ANSI-accredited standard that addresses the design of workstations, furniture, and computer systems.

The BIFMA G1 guidelines are published by the Business and Institutional Furniture Manufacturers Association (BIFMA), which is a nonprofit trade association that develops, maintains, and publishes safety and performance standards for furniture products. BIFMA G1, *Ergonomics Guideline for Furniture Used in Office Work Spaces Designed for Computer Use,* is a set of guidelines that provides dimensional guidance to office furniture manufacturers in the development, design, and specification of ergonomic solutions for computer work spaces.

DESK HEIGHT FLEXIBILITY

Courtesy of Knoll, Inc.

Figure 10-6. Part 2: Desk Height Flexibility of Feature 73, Ergonomics: Visual and Physical, requires that at least 30% of workstations have the ability to alternate between sitting and standing positions.

BY THE NUMBERS

- *In 2010, nearly 7% (more than 169 million) of all disability-adjusted life years (DALYs) worldwide resulted from musculoskeletal disorders.*

FEATURE 74, EXTERIOR NOISE INTRUSION

Feature 74, Exterior Noise Intrusion, is a precondition for the Core and Shell and New and Existing Buildings project types, but it is an optimization for the New and Existing Interiors type. Prolonged exposure to excessive levels of exterior noise can increase the stress levels of building occupants as well as the risk of health problems. These health problems can include hypertension, diabetes, strokes, and heart attacks. Vehicle traffic, low-flying aircraft, construction, and human activity can create excessive exterior noise. **See Figure 10-8.**

Part 1: Sound Pressure Level

Sound pressure level, also known as acoustic pressure, is the pressure variation associated with sound waves, usually measured in decibels (dB). A *decibel (dB)* is a unit of measurement for sound that is a logarithmic unit in which an increase in 10 decibels equals an increase by a factor of 10. A-weighted decibels (dBA) are used for Part 1 because the frequency response is similar to that of the human ear and is an accurate measure of how sound is actually perceived. The dBA reading on a sound meter disregards the low and high frequencies most humans cannot hear.

SEAT FLEXIBILITY

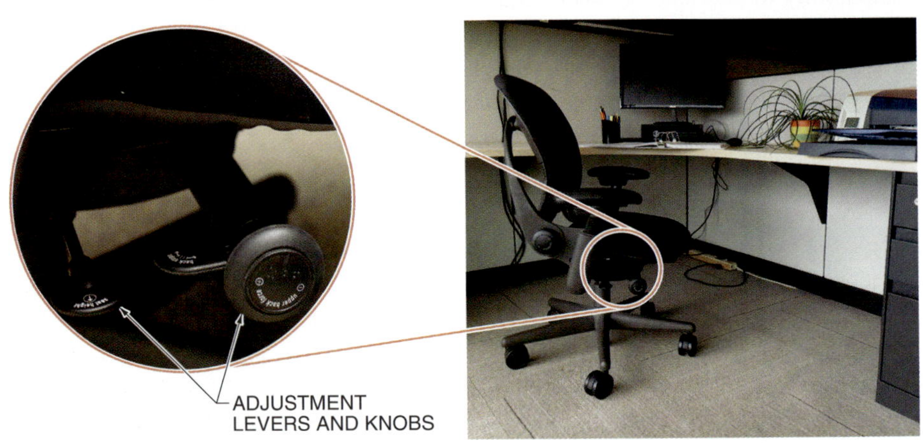

Figure 10-7. Adjustable chairs allow occupants to change the height and depth of their seats so that they can find better ergonomic comfort at their workstations.

EXTERIOR NOISE INTRUSION

Figure 10-8. Vehicle traffic, low-flying aircraft, construction, or human activity can create excessive exterior noise that increases the stress levels of building occupants.

Part 1: Sound Pressure Level of Feature 74 provides project teams with a limit on the amount of exterior noise that is allowed to infiltrate regularly occupied spaces. The sound should be measured when the regularly occupied spaces and adjacent spaces are empty and within 1 hour of normal business hours. The maximum average sound pressure level from exterior noise intrusion for any occupied space cannot exceed 50 dBA for the requirements of this feature.

FEATURE 75, INTERNALLY GENERATED NOISE

Feature 75, Internally Generated Noise, is a precondition for the New and Existing Interiors and New and Existing Buildings project types. The second part is also an optimization for Core and Shell projects. Interior noise from occupants, office equipment, HVAC and mechanical systems, and electronics can cause distractions, a decreased sense of privacy, and lower productivity.

Part 1: Acoustic Planning

Part 1: Acoustic Planning of Feature 75 requires that an acoustic plan be developed for a project. The acoustic plan must identify loud and quiet zones within the project as well as any noise-producing equipment in the spaces. **See Figure 10-9.** The acoustic plan is then used to configure the spaces to reduce the amount of interior noise.

Part 2: Mechanical Equipment Sound Levels

Part 2: Mechanical Equipment Sound Levels of Feature 75 requires that specific occupied spaces not exceed maximum noise criteria (NC). *Noise criteria (NC)* is the sound pressure limits of the octave band spectra ranging from 63 Hz to 8000 Hz. **See Figure 10-10.**

Noise criteria is used instead of decibels for measuring sound in occupied building spaces because it more accurately measures the low-frequency noise produced by HVAC and other mechanical equipment. Low-frequency noise can cause a lot of disruption for occupants in certain spaces, leading to reduced concentration and mental performance. After an interior build-out is completed, the HVAC and mechanical systems must meet the following NC levels for the listed occupied spaces:

- open office spaces and lobbies that are regularly occupied and/or contain workstations—maximum NC of 40
- enclosed offices—maximum NC of 35
- conference rooms and breakout rooms—maximum NC of 30 (25 recommended)

ACOUSTIC PLANNING

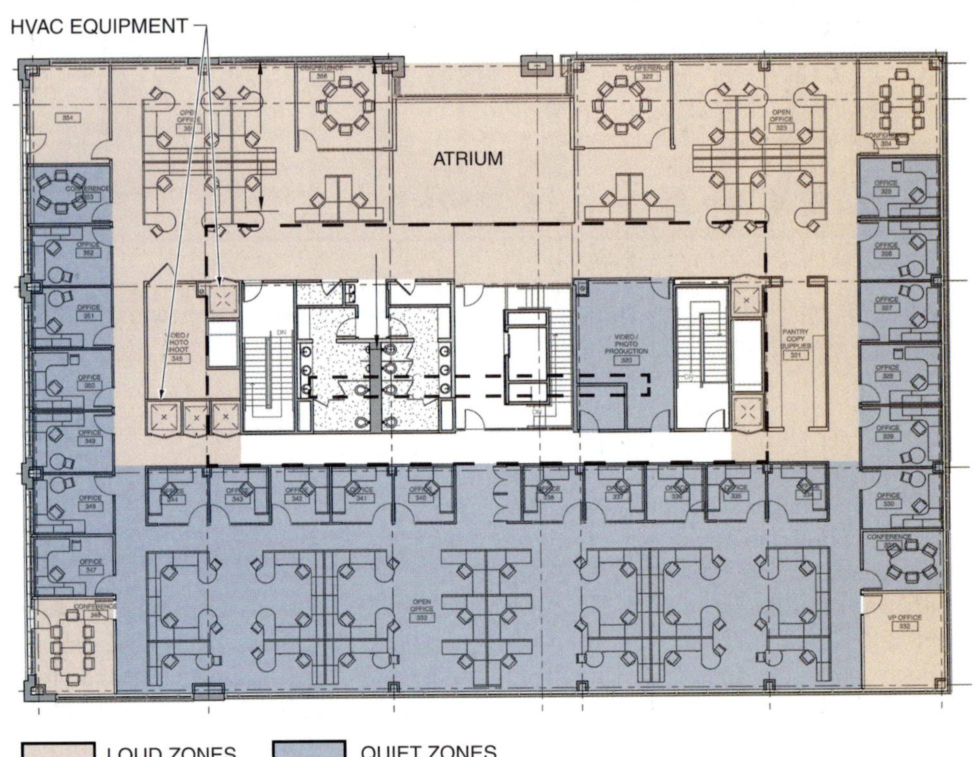

Figure 10-9. Part 1: Acoustic Planning of Feature 75, Internally Generated Noise, requires that an acoustic plan be developed to identify loud and quiet zones within a project as well as any noise-producing equipment in the spaces.

NOISE CRITERIA (NC) CURVES

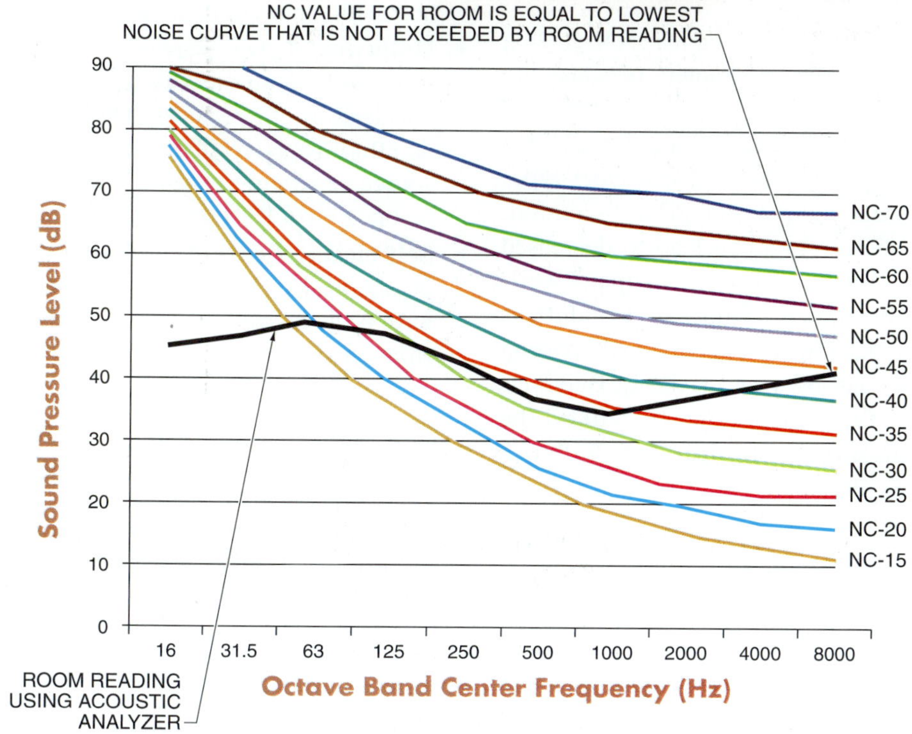

Figure 10-10. The low-frequency noise produced by HVAC systems and other equipment is more accurately measured by noise criteria (NC) than decibels.

FEATURE 76, THERMAL COMFORT

Thermal comfort is the condition of satisfaction with the thermal environment. Feature 76, Thermal Comfort, is a precondition for all three project types that requires that a sufficient level of thermal comfort be provided to building occupants. ASHRAE Standard 55-2013, *Thermal Environmental Conditions for Human Occupancy*, provides criteria for both mechanically ventilated and naturally conditioned spaces, which must be met for this feature, depending on the type of ventilation system used for each space in the building.

BY THE NUMBERS

- *In 2006, only 11% of the office buildings surveyed in the United States provided thermal environments that met generally accepted goals of occupant satisfaction.*

Part 1: Ventilated Thermal Environment

Part 1: Ventilated Thermal Environment of Feature 76 requires that all the spaces in a mechanically ventilated building meet the design, operating, and performance criteria of ASHRAE Standard 55-2013 Section 5.3, *Standard Comfort Zone Compliance*. A mechanically ventilated building depends on a system of fans, heating units, and chillers to provide heating and cooling to occupied spaces.

Part 2: Natural Thermal Adaptation

Part 2: Natural Thermal Adaptation of Feature 76 requires that all the spaces in a naturally conditioned building meet the criteria of ASHRAE Standard 55-2013 Section 5.4, *Adaptive Comfort Model*. A naturally conditioned building depends on pressure differences to create airflow and provide heating and cooling to the spaces in the building.

FEATURE 77, OLFACTORY COMFORT

Odors from cleaning chemicals, off-gassing materials, copiers and printers, personal care items such as perfumes and shampoos, and food sources can cause headaches, eye and nasal discomfort, nausea, and other types of physical and psychological discomfort. These discomforts can lead to decreased productivity and lower occupant well-being. Feature 77, Olfactory Comfort, is an optimization for the New and Existing Interiors and New and Existing Buildings project types that aims to reduce the transmission of strong odors within a building space.

Part 1: Source Separation

Part 1: Source Separation of Feature 77 requires that all restrooms, janitorial closets, kitchens, cafeterias, and pantries use source separation methods to prevent strong odors from migrating to other occupied spaces. A project must use one or more of the following methods:

- **Negative pressurization.** Exhaust fans are used to create negative pressurization by reducing the air pressure in a space to a lower level than the surrounding spaces. The pressure differential keeps air and odors from migrating to the surrounding higher-pressure spaces. Negative pressurization is a highly effective method of odor control, but it may be considerably more expensive than other strategies due to the increased mechanical system requirements and energy consumption.
- **Interstitial rooms.** Interstitial rooms are unoccupied spaces between regularly occupied spaces. These rooms may be used to house mechanical systems or other utility support, or they may serve no purpose other than separating two or more occupied spaces.
- **Vestibules.** Vestibules are small areas that connect two or more sets of doorways and act as airlocks to prevent free airflow between spaces. Vestibules are usually found between the main entrances of buildings, but they may be found connecting interior doorways as well.
- **Hallways.** Hallways are common areas used to provide access to doorways that lead to other spaces. They are not as effective as vestibules at providing source separation because they tend to be open on at least one end.
- **Self-closing doors.** Self-closing doors automatically close the doors after they have been opened. This prevents airflow from one space to another.

FEATURE 78, REVERBERATION TIME

Reverberation time is the time it takes for sound to decay, expressed in seconds. **See Figure 10-11.** The decay rate depends on the sound-absorptive properties of the walls, ceilings, floors, and furnishings in a room, the room's geometry, and the frequency of the sound. Long reverberation times can cause excessive noise levels to persist and lead to increased stress, difficulty in communication, and loss of productivity. Feature 78, Reverberation Time, is an optimization for the New and Existing Interiors and New and Existing Buildings project types that sets maximum limits for reverberation time to maintain comfortable sound levels.

A source separation method, such as a vestibule, must be used to prevent the migration of strong odors from space to space.

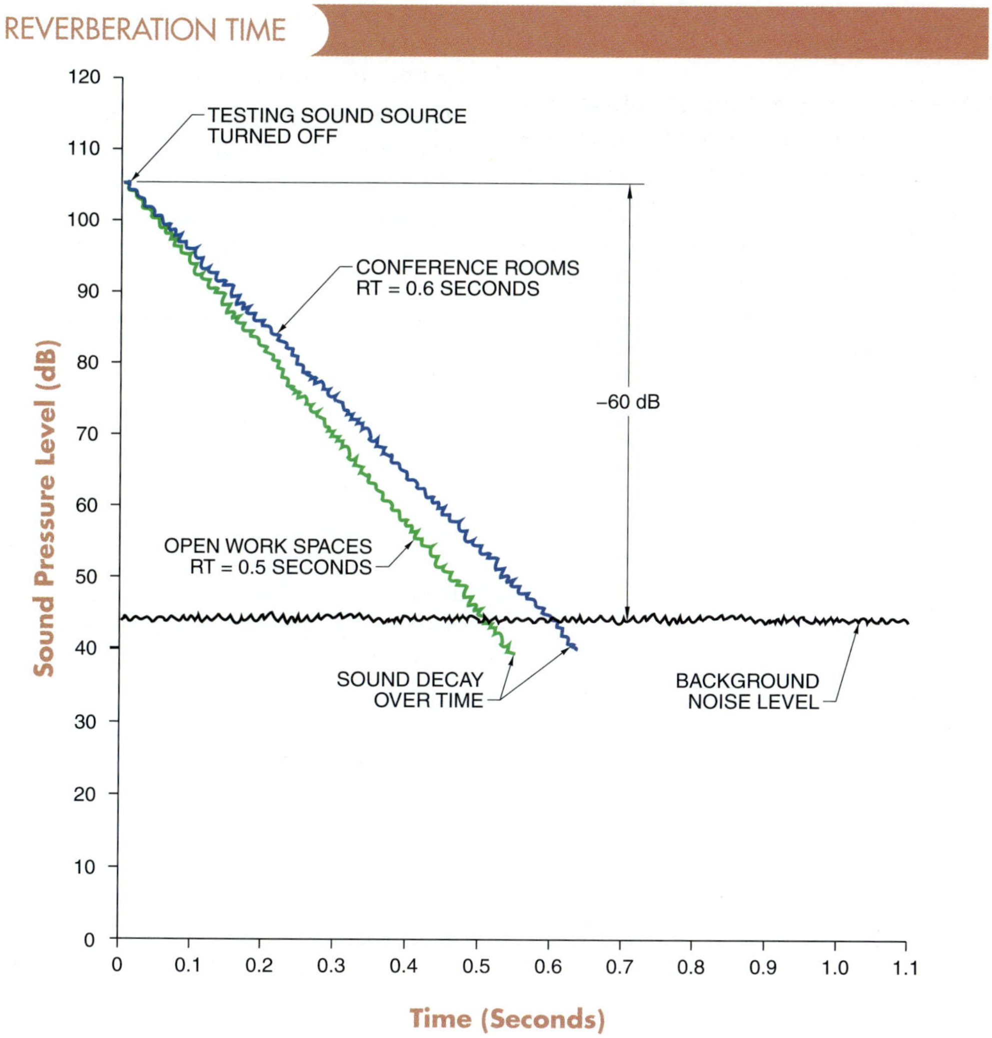

Figure 10-11. Reverberation time, which is the time it takes for sound to decay, can cause excessive noise levels to persist and lead to increased stress, difficulty in communication, and loss of productivity if it is too long.

Part 1: Reverberation Time

Part 1: Reverberation Time of Feature 78 uses a level of reverberation time called RT60, which is the metric that describes the length of time taken for a sound to decay by 60 dB from its original level. Part 1 limits the maximum RT60 for conference rooms and open work spaces. For conference rooms, the maximum RT60 is 0.6 seconds, meaning it must take 0.6 seconds or less for a sound to decay 60 dB from its original level. For open work spaces, the maximum RT60 is 0.5 seconds.

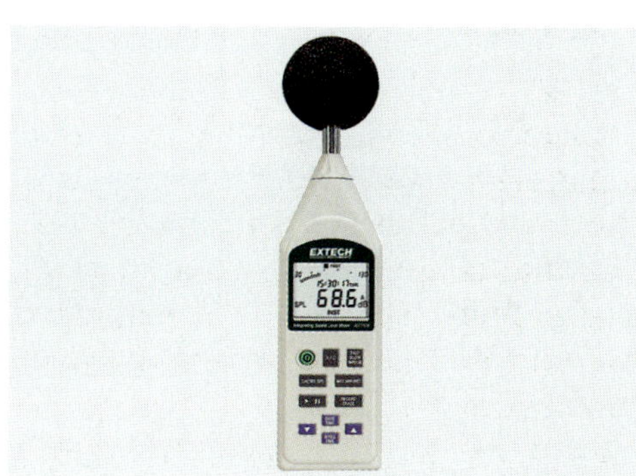

Extech Instruments

A sound-level meter with a timer can be used to measure RT60, or how long it takes sound to decay by 60 dB in an enclosed space.

FEATURE 79, SOUND MASKING

Ambient silence occurs when a space is significantly quiet, making any small noise disturbing and reducing the sense of privacy. *Sound masking* is the use of electronic devices to generate low-level background noise in order to mask sound distractions. Sound masking allows for increased confidentiality in communications and reduced distractions from unwanted and unnecessary sounds. Feature 79, Sound Masking, is an optimization for the New and Existing Interiors and New and Existing Buildings project types that aims to reduce the sound distractions accompanying ambient silence.

Part 1: Sound Masking Use

Part 1: Sound Masking Use of Feature 79 requires a sound masking system to be used for all open office work spaces. For large areas of office space, a central control module and multiple emitters can be used for sound masking. Sound masking can also be provided at individual workstations or offices using individual controllers and speakers. **See Figure 10-12.**

WELLNESS FACT

Sound masking can be accomplished in various ways, such as electronics in air plenums or even small tabletop fountains for personal spaces.

Part 2: Sound Masking Limits

Part 2: Sound Masking Limits of Feature 79 sets limits for the sound masking systems required by Part 1. For open work spaces, the levels of sound masking systems should fall between 45 dBA to 48 dBA when measured from the nearest work space. For enclosed offices, the levels of sound masking systems should fall between 40 dBA to 42 dBA when measured from the nearest work space.

FEATURE 80, SOUND REDUCING SURFACES

Feature 80, Sound Reducing Surfaces, is an optimization for the New and Existing Interiors and New and Existing Buildings project types. Sound-reducing surfaces are used to counteract excessive sound transmission that might be present after a building is designed and built. Examples of sound-reducing surfaces include acoustical ceiling tiles, ceiling baffles, or decorative acoustic wall panels.

Sound-reducing surfaces can be placed on ceilings and walls to meet the requirements of this feature. The sound-reducing surfaces must help a space meet a minimum noise reduction coefficient (NRC) for a specified percentage of surface area for that space. *Noise reduction coefficient (NRC)* is the average value that determines the absorptive properties of materials.

SOUND MASKING

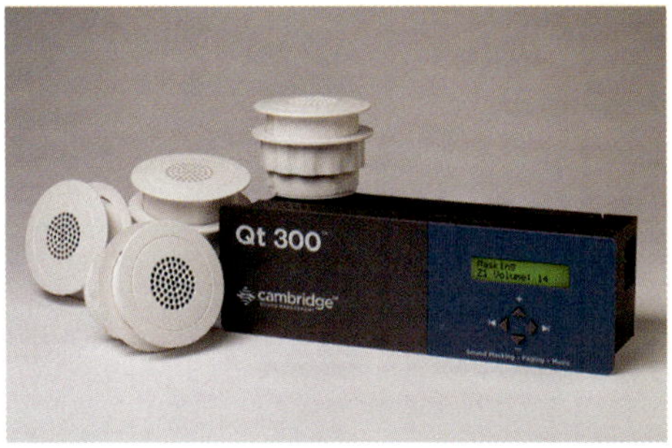

FOR LARGE AREAS

AT INDIVIDUAL WORKSTATIONS

Steelcase

Figure 10-12. Sound masking is the use of electronic devices in large areas or at individual workstations to generate low-level background noise in order to mask sound distractions.

An NRC ranges from 0 to 1.00. **See Figure 10-13.** An NRC of 0 indicates absolute sound reflection. An NRC of 1.00 indicates absolute sound absorption. For example, an acoustical ceiling tile with an NRC value of 0.65 would absorb 65% of the sound that strikes the surface of the tile while reflecting back 35%. NRC values are provided by the product manufacturer.

WELLNESS FACT

The U.S. General Services Administration (GSA) explains that three-dimensional, cloud-like shapes or flaglike materials can be used like baffles to reduce the transmission of sound. These sound reduction methods not only reduce acoustical dissatisfaction but also create visual interest in a space.

Part 1: Ceilings

Part 1: Ceilings of Feature 80 addresses ceilings, the largest reflective surface in a space. Open work spaces have a greater need for ceiling-surface sound reduction than conference or teleconference rooms. Part 1 requires open work spaces to have a minimum NRC of 0.9 for the entire surface area of the ceiling, excluding lights, skylights, diffusers, and grilles. Conference and teleconference rooms are required to have a minimum NRC of 0.8 for at least 50% of the surface area of the ceiling, excluding lights, skylights, diffusers and grilles.

Part 2: Vertical Surfaces

Part 2: Vertical Surfaces of Feature 80 addresses the vertical surfaces in conference and teleconference rooms, open work spaces, and cubicle partitions. Part 2 requires that enclosed offices, conference rooms, and teleconference rooms have a minimum NRC of 0.8 for at least 25% of the surface area of the surrounding walls. Open work spaces must have a minimum NRC of 0.8 for at least 25% of the surface area of the surrounding walls. Partitioned office spaces must be at least 1.2 m (48″) high and have a minimum NRC of 0.8.

Average Noise Reduction Coefficients for Building Materials

Building Materials	Noise Reduction Coefficient (NRC)
Acoustical Ceiling Tile	0.50 to 0.90
Brick	0.02 to 0.05
Carpet	0.30 to 0.55
Concrete	0.05 to 0.20
Cork Wall Tile	0.70
Glass	0.05 to 0.10
Gypsum Board	0.05
Interior Furnishings	0.30 to 0.60
Marble or Terrazzo	0.00
Polyurethane Foam (Open Cell)—25 mm (1″)	0.30
Semi-Rigid Fiberglass—25 mm (1″)	0.75
Sprayed Cellulose Fiber—25 mm (1″)	0.75
Steel	0.00 to 0.10
Wood	0.05 to 0.15

Figure 10-13. The NRC values of common building materials can vary greatly based on the sound absorptive properties of the material.

FEATURE 81, SOUND BARRIERS

Sound waves travel through air and solid materials. However, walls and doors with unfilled gaps or missing insulation can allow increased levels of unwanted sound to pass through to an adjacent room. Sound transmission to adjacent rooms can be a distraction to building occupants. Best practices in sound control design and building construction can reduce the amount of sound transmission. Feature 81, Sound Barriers, is an optimization for the New and Existing Interiors and New and Existing Buildings project types that addresses wall construction specifications and methodology as well as doorway specifications for reduced sound transmission.

Part 1: Wall Construction Specifications

Part 1: Wall Construction Specifications of Feature 81 requires that interior partition walls of occupied spaces, such as enclosed offices, conference rooms, and teleconference rooms, meet minimum noise-isolation- class levels.

Noise isolation class (NIC) is a field test for determining the sound transmitting abilities of a wall. Higher NIC values indicate more effective sound cancellation between spaces. Because NIC is a field test, it considers the full environment in which the wall is placed, not just the wall itself. This provides a more accurate measurement of sound cancellation compared to sound-transmission- class levels. *Sound transmission class (STC)* is a laboratory method for determining the sound transmission through a wall.

When a sound masking system is used in enclosed offices, the interior partition walls must meet a minimum NIC of 35. When no sound masking system is being used, the interior partition walls of enclosed offices must meet a minimum NIC of 40. Because sound levels in conference rooms and teleconference rooms can be higher than other spaces, interior partition walls that adjoin private offices or other conference or teleconference rooms require a greater NIC level of 53.

Part 2: Doorway Specifications

Doorways may act as a path for sound to travel between adjacent spaces. Part 2: Doorway Specifications of Feature 81 requires that at least one of three sound reduction strategies be used to reduce the amount of noise transfer through doorways. **See Figure 10-14.** Doors connecting to private offices, conference rooms, and teleconference rooms must be constructed with at least one of the following strategies:

- **Gaskets.** Gaskets are strips of rubber or foam that are placed on the door-side of the door stop and create an airtight seal against the door slab. The airtight seal does not allow sound to leak through.
- **Sweeps.** Sweeps are attached to the bottom of a door slab and seal the space between the bottom of the door and the floor when the door is closed.
- **Nonhollow cores.** Non-hollow-core door slabs are composed of solid wood or another material such as particleboard or medium-density fiberboard (MDF). The increased density of these door slabs reduces the sound transmission properties of the doorway.

WELLNESS FACT

STC ratings for building assemblies do not signify how much sound is blocked but rather the level of sound transmitted through the assemblies. For example, an assembly with an STC of 40 reduces loud speech down to a murmur on its other side.

Part 3: Wall Construction Methodology

Part 3: Wall Construction Methodology of Feature 81 addresses best practices for wall construction to reduce the amount of sound transmission that can occur between occupied spaces. **See Figure 10-15.** Part 3 requires the following strategies for all interior walls enclosing regularly occupied spaces:

- properly sealing all acoustically rated partitions at the top and bottom tracks
- staggering all gypsum board seams
- packing and sealing all penetrations through the wall

SOUND REDUCTION STRATEGIES FOR DOORS

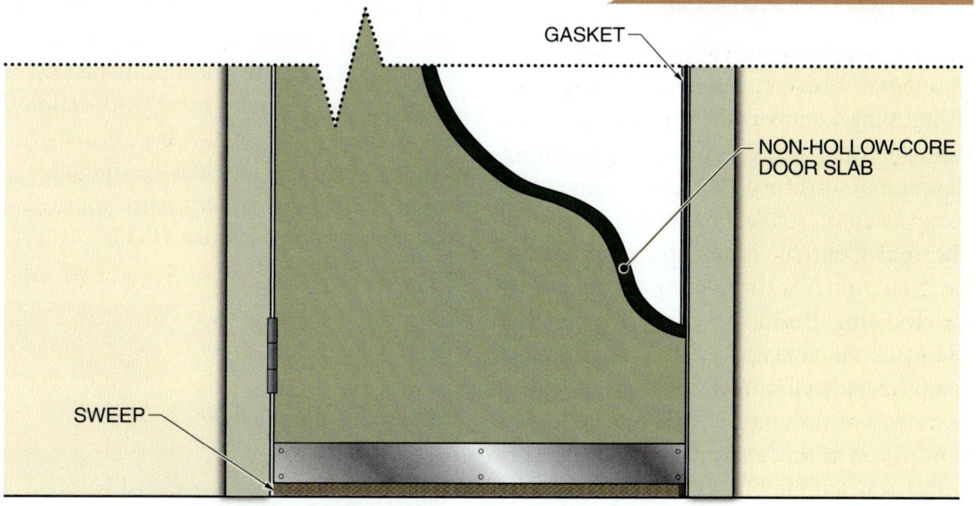

Figure 10-14. Three doorway strategies that can be used to reduce the amount of noise transfer include gaskets, sweeps, and nonhollow cores.

WALL CONSTRUCTION METHODOLOGY

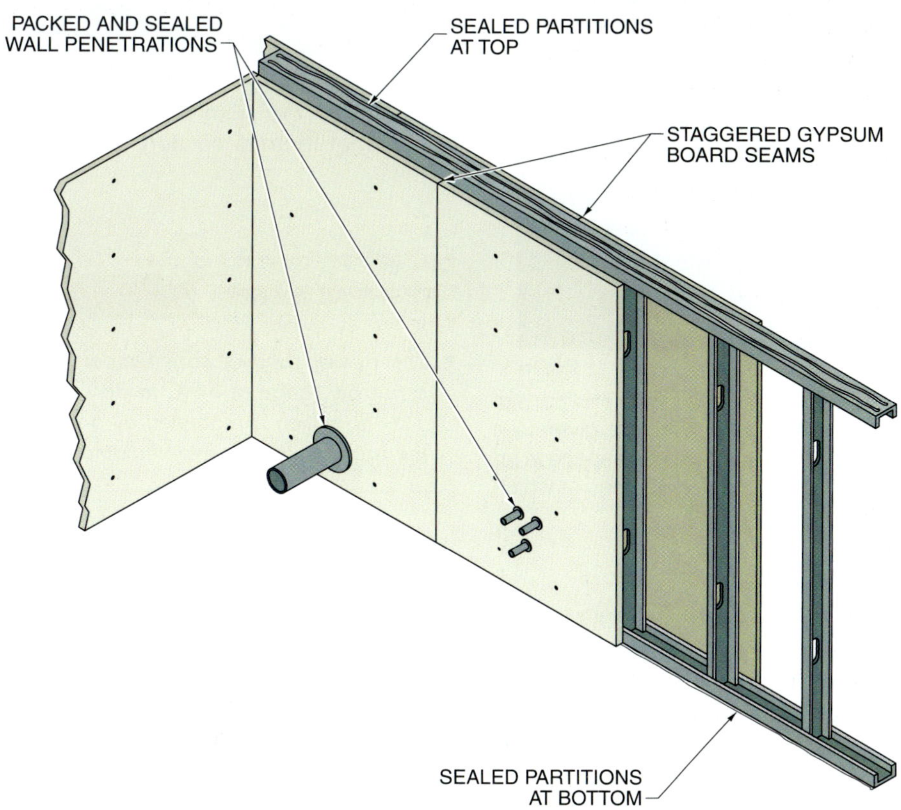

Figure 10-15. Part 3: Wall Construction Methodology of Feature 81, Sound Barriers, addresses best practices in wall construction to reduce the amount of sound transmission that can occur between regularly occupied spaces.

FEATURE 82, INDIVIDUAL THERMAL CONTROL

In some buildings, HVAC zoning is used as an attempt to make spaces comfortable for all building occupants. However, different metabolisms, body types, and clothing choices make it difficult to keep every building occupant comfortable. Feature 82, Individual Thermal Control, is an optimization for the New and Existing Interiors and New and Existing Buildings project types that addresses the concepts of free address and personal thermal comfort. *Free address* is the ability of occupants to choose their own work space within the office or workplace. In this way, building occupants can find the work area in which they are most comfortable. **See Figure 10-16.**

FREE ADDRESS

Steelcase

Figure 10-16. Occupants who are given free address to choose their own work space in an office building may find workstations at which they are more comfortable.

Part 1: Free Address

Part 1: Free Address of Feature 82 applies to projects over 200 m^2 (2150 ft^2). Part 1a requires that a building provide a thermal gradient of at least 3°C (5°F) across open work spaces and between floors or in rooms with more than 10 people. Part 1b requires that all open office spaces with occupants performing tasks that use similar workstations allow at least 50% free address so occupants can select a work space with a desired temperature.

Part 2: Personal Thermal Comfort Devices

Part 2: Personal Thermal Comfort Devices of Feature 82 requires that occupants in work spaces containing 10 or more workstations in the same HVAC zone have access to thermal comfort devices. These devices include fans and heated/cooled office chairs but exclude space heaters. **See Figure 10-17.**

PERSONAL THERMAL COMFORT DEVICES

Figure 10-17. Occupants must have access to personal thermal comfort devices such as fans.

BY THE NUMBERS

- *The human body's core temperature has a narrow range of 36°C to 38°C (97°F to 100°F), which is regulated by a portion of the brain called the hypothalamus.*

FEATURE 83, RADIANT THERMAL COMFORT

The first part of Feature 83, Radiant Thermal Comfort, is an optimization for all three project types. The second part is an optimization only for the New and Existing Interiors and New and Existing Buildings types. This feature encourages the use of radiant temperature systems to increase energy efficiency and thermal comfort. Radiant temperature systems are less likely to distribute allergens than forced-air systems.

Radiant temperature systems that can be used for this feature include hydronic heating and/or cooling systems and electric radiant systems. A *hydronic radiant heating and/or cooling system* is a radiant temperature system that uses water or another heat-transfer fluid to carry heated or chilled water from the point of generation to the point of use. The point of use can be underfloor or in-slab piping, ceiling or wall panels, or another device that transfers heat. **See Figure 10-18.** An *electric radiant system* is a radiant temperature system that uses the heat created by the resistance of wiring to electrical current in order to warm the air or material around it.

RADIANT THERMAL COMFORT

Legend Valve and Fitting, Inc.

Figure 10-18. A hydronic radiant heating and/or cooling system uses water or another heat-transfer fluid to carry heated or chilled water from the point of generation to the point of use.

Part 1: Lobbies and Other Common Spaces

Part 1: Lobbies and Other Common Spaces of Feature 83 requires that all lobbies and other common spaces meet the requirements of ASHRAE Standard 55-2013 through the use of either hydronic radiant heating and/or cooling systems or electric radiant systems. Other common spaces addressed by Part 1 include elevators, stairs, atriums, and restrooms.

Part 2: Offices and Other Regularly Occupied Spaces

Part 2: Offices and Other Regularly Occupied Spaces of Feature 83 requires that at least 50% of the floor area in all offices and other regularly occupied spaces meet the requirements of ASHRAE Standard 55-2013. Either hydronic radiant heating and/or cooling systems or electric radiant systems can be used to meet this requirement.

KEY TERMS AND DEFINITIONS

acoustics: The study of sound and the properties of a space or building that determine how sound is transmitted or reflected within it or through it.

American with Disabilities Act (ADA): An enacted law that prohibits discrimination against persons with disabilities and ensures equal opportunity in employment, state and local government services, public accommodations, commercial facilities, and transportation.

decibel (dB): A unit of measurement for sound that is a logarithmic unit in which an increase in 10 decibels equals an increase by a factor of 10.

electric radiant system: A radiant temperature system that uses the heat created by the resistance of wiring to electrical current in order to warm the air or material around it.

ergonomics: The science of adapting objects, spaces, and processes, such as workstations and workflow, to accommodate people's capabilities in a safe and efficient manner.

free address: The ability of occupants to choose their own work space within the office or workplace.

hydronic heating and/or cooling system: A radiant temperature system that uses water or another heat-transfer fluid to carry heated or chilled water from the point of generation to the point of use.

KEY TERMS AND DEFINITIONS *(continued)*

International Organization for Standardization (ISO): An independent, nongovernmental international organization that develops consensus-based, market-relevant standards for worldwide use.

musculoskeletal disorder: An injury or condition that affects the body's movements or its muscles, bones, tendons, nerves, or ligaments.

noise criteria (NC): The sound pressure limits of the octave band spectra ranging from 63 Hz to 8000 Hz.

noise insulation class (NIC): A field test for determining the sound transmitting abilities of a wall.

noise reduction coefficient (NRC): The average value that determines the absorptive properties of materials.

olfaction: The sense of smell.

olfactory comfort: A person's perception of air quality within an environment based on their sense of smell.

psychrometric chart: A graphical representation of the properties of air at various conditions.

psychrometrics: The mathematical and graphical study of the properties of air.

reverberation time: The time it takes for sound to decay, expressed in seconds.

sound: Energy composed of pressure waves, or vibrations, that travel through the air or another material and produce an audible signal when they reach a person's ear.

sound masking: The use of electronic devices to generate low-level background noise in order to mask sound distractions.

sound pressure level: The pressure variation associated with sound waves, usually measured in decibels (dB). Also known as acoustic pressure.

sound transmission class (STC): A laboratory method for determining the sound transmission through a wall.

thermal comfort: The condition of satisfaction with the thermal environment.

CHAPTER REVIEW

Completion

______________ **1.** ___ is the study of sound and the properties of a space or building that determine how sound is transmitted or reflected within it or through it.

______________ **2.** Common ___ that can occur in an office building include carpal tunnel syndrome in the hands and arms, tendonitis in the hands and wrists, and pain in the back, hips, or neck.

______________ **3.** ___ comfort is a person's perception of air quality within an environment based on their sense of smell.

______________ **4.** The standard used for thermal comfort in WELL is ASHRAE Standard ___, *Thermal Environmental Conditions for Human Occupancy.*

______________ **5.** Feature 72, ___, is a precondition for all three project types that aims to ensure that projects provide an equitable built environment.

______________ **6.** Part 1: Sound Pressure Level of Feature 74, Exterior Noise Intrusion, requires that the average sound pressure level from exterior noise intrusion not exceed ___ dBA.

CHAPTER REVIEW *(continued)*

_______________ **7.** ___ is the sound pressure limits of the octave band spectra ranging from 63 Hz to 8000 Hz.

_______________ **8.** Part 1: Reverberation Time of Feature 78, Reverberation Time, uses a level of reverberation time called ___, which is the metric that describes the length of time taken for a sound to decay by 60 dB from its original level.

_______________ **9.** Feature 79, ___, calls for the use of electronic devices to generate low-level background noise to allow for increased confidentiality in communications and reduced distractions from unwanted and unnecessary sounds.

_______________ **10.** The ceilings of open work spaces must have a minimum NRC of ___ for their entire surface area, while the ceilings of conference and teleconference rooms must have a minimum NRC of ___ for at least 50% of their surface area to meet the requirements of Part 1: Ceilings of Feature 80, Sound Reducing Surfaces.

_______________ **11.** Gaskets, sweeps, and non-hollow cores are the three sound reduction strategies that can be used to meet the requirements of Part 2: ___ of Feature 81, Sound Barriers.

_______________ **12.** ___ is the ability of occupants to choose their own work spaces within the office or workplace.

_______________ **13.** Part 1: Free Address of Feature 82, Individual Thermal Control, applies to projects with areas over ___.

_______________ **14.** Part 2: Personal Thermal Comfort Devices of Feature 82, Individual Thermal Control, requires that occupants in work spaces containing ___ or more workstations in the same HVAC zone have access to low-power thermal comfort devices.

Short Answer

1. List the negative effects that unpleasant or overwhelming smells can have on a person.

2. List the six factors of thermal comfort per ASHRAE Standard 55-2013.

3. List the three parts of Feature 73, Ergonomics: Visual and Physical, and their requirements.

CHAPTER REVIEW *(continued)*

4. List the maximum noise criteria (NC) levels for open office spaces and lobbies, enclosed offices, and conference and breakout rooms that meet the requirements of Part 2: Mechanical Equipment Sound Levels of Feature 75, Internally Generated Noise.

5. Explain the difference between a mechanically ventilated building and a naturally conditioned building.

6. List the five source separation methods that can be used to prevent strong odors from migrating to other occupied spaces.

7. What is the difference between noise isolation class (NIC) and sound transmission class (STC)?

8. Describe the two radiant temperature systems that can be used to meet the requirements of Feature 83, Radiant Thermal Comfort.

WELL AP EXAM PRACTICE QUESTIONS

1. The Comfort Concept of WELL Building Standard promotes ergonomic solutions for which health issue that is responsible for 380,600 missed work days in the United States (about one-third of total missed worked days)?

A. Obesity
B. Musculoskeletal disorders
C. Addiction
D. Major depressive disorders

2. According to Part 1: Ventilated Thermal Environment of Feature 76, Thermal Comfort, all spaces in a mechanically ventilated building must comply with which standard?

A. ASHRAE Standard 55-2013 Section 5.3, Standard Comfort Zone Compliance
B. ASHRAE Standard 55-2013 Section 5.4, Adaptive Comfort Model
C. ASHRAE Standard 62.1-2013
D. HFES 100 standard

3. Which feature in the Comfort Concept does a project meet if the sound in its open work spaces decays by 60 dB in 0.45 seconds?

A. Feature 74, Exterior Noise Intrusion
B. Feature 75, Internally Generated Noise
C. Feature 78, Reverberation Time
D. Feature 80, Sound Reducing Surfaces

4. What is the minimum height and NRC value for the partitions of office spaces required by Part 2: Vertical Surfaces of Feature 80, Sound Reducing Surfaces?

A. Partitions must reach at least 1 m (40″) and have a minimum NRC of 0.7.
B. Partitions must reach at least 1.2 m (48″) and have a minimum NRC of 0.7.
C. Partitions must reach at least 1.2 m (48″) and have a minimum NRC of 0.8.
D. Partitions must reach at least 1.5 m (60″) and have a minimum NRC of 0.8.

5. What type of annotated document is required for Part 1: Free Address of Feature 82, Individual Thermal Control?

A. Policy document
B. Architectural drawing
C. Operations schedule
D. Professional narrative

CHAPTER ELEVEN

MIND AND INNOVATION

Mental and physical health are closely linked, and a change in one can impact the other. A healthy mental state provides a significant boost to a person's psychological and physical well-being. The built environment, especially as a workplace, can play a meaningful role in maintaining or improving a person's mental health. The goal of the Mind Concept in the WELL Building Standard is to improve building occupants' mental health by presenting features that help relieve stress, improve sleep quality, improve overall mood, and make the occupants feel welcome in the building spaces. The goal of the Innovation features is to promote novel strategies for addressing human health and wellness.

OBJECTIVES

- Identify the features in the Mind Concept as preconditions or optimizations.
- Describe the benefits of a healthy mental state as well as the negative consequences of poor mental health.
- Explain the integrative design process and the role of charrettes within it.
- Describe the strategies used by the features in the Mind Concept to reduce the environmental and psychosocial stress levels of building occupants.
- Explain how a comprehensive health benefits plan and a workplace family support plan promote well-being.
- Explain the importance of aesthetically beautiful building elements and how they can be incorporated into the building design.
- Explain biophilia and how biophilic design elements can be incorporated into the building design.
- Describe the strategies and policies that employers can adopt to promote healthy sleep hygiene and behaviors.
- Describe how building spaces can be designed to reduce distractions while also promoting privacy and collaboration between building occupants.
- Explain how project teams can work innovative features that do not yet appear in the WELL Concepts into their projects.

KEY TERMS

- altruism
- biomimicry
- biophilia
- charrette
- collaboration zone
- Declare label
- employee assistance program (EAP)
- G4 Sustainability Reporting Guidelines
- health literacy
- Health Product Declaration (HPD)
- integrative design process
- JUST Program
- organizational transparency
- sleep hygiene
- spatial familiarity

Learner Resources

ATPeResources.com/QuickLinks
Access Code: 935534

MIND CONCEPT

Overall positive health and wellness is dependent on the combination of mental and physical well-being. A decline in either mental or physical well-being can have a significant impact on the other. The mental well-being of building occupants is therefore very important to their overall state of health. WELL aims to positively influence building occupants' mood, quality of sleep, stress levels, and overall mental well-being.

The Mind Concept in WELL includes 17 features that promote strategies and policies such as beauty through design, biophilia, healthy sleep, material and organizational transparency, and altruism. Core and Shell projects must meet three preconditions, but four optimizations can be pursued for higher certification. New and Existing Interiors and New and Existing Buildings must meet five preconditions, but 12 optimizations can be pursued for higher certification. **See Figure 11-1.**

MENTAL HEALTH

Mental health is the state of a person's emotional, psychological, and social well-being. A healthy state of mind can have a positive impact on personal and work relationships, which leads to higher productivity, better performance, less absenteeism, and fewer workplace accidents. Businesses benefit by supporting the mental health of workers through lower healthcare costs, increased quality and productivity, and employee retention. Untreated mental health diseases and mood disorders cost businesses billions of dollars every year in lost economic output.

When a person has poor mental health, mood disorders such as depression, stress, and anxiety can lead to personal health problems and problems within the workplace. Depression can suppress the function of the immune system and may lead to an increased risk of heart disease. Stress and anxiety can lead to an increased risk of metabolic syndrome, cardiovascular disease, gastrointestinal disorders, and negative skin conditions. The problems that mood disorders can cause within the workplace include decreased productivity, absenteeism, and negative interactions between employees.

FEATURE 84, HEALTH AND WELLNESS AWARENESS

Wellness education, or health literacy, is an important part of building occupant health. *Health literacy* is the ability for people to obtain, study, and understand the health information and available health services that are communicated to them through health literature and presented options. When people are fully informed and aware of health-related options, they can make educated lifestyle and wellness decisions.

Feature 84, Health and Wellness Awareness, is a precondition for all three project types. This feature encourages health literacy by promoting the availability of health and wellness literature, educational resources, and information on the features and benefits of the WELL Building Standard.

Denmarsh Photography, Inc.

The Mind Concept promotes strategies such as beauty through design and biophilia.

Part 1: WELL Building Standard® Guide

Part 1: WELL Building Standard® Guide of Feature 84 requires that a guide describing the WELL features achieved by the project be available to all building occupants. A guide allows occupants to more clearly understand the health and wellness benefits of the project's features, including benefits that extend beyond the built environment.

WELL Building Standard Features: Mind Concept and Innovation...						
Features	Project Type			Verification Documentation		
	Core and Shell	New and Existing Interiors	New and Existing Buildings	Letter of Assurance	Annotated Documents	On-Site Assessment
84, Health and Wellness Awareness						
Part 1: WELL Building Standard® Guide	P	P	P			Visual Inspection
Part 2: Health and Wellness Library	P	P	P			Visual Inspection
85, Integrative Design						
Part 1: Stakeholder Charrette	P	P	P		Policy Document	
Part 2: Development Plan	P	P	P		Policy Document	
Part 3: Stakeholder Orientation	P	P	P		Policy Document	
86, Post-Occupancy Surveys						
Part 1: Occupant Survey Content	–	P	P		Policy Document	
Part 2: Information Reporting	–	P	P		Policy Document	
87, Beauty and Design I						
Part 1: Beauty and Mindful Design	P	P	P		Professional Narrative	
88, Biophilia I–Qualitative						
Part 1: Nature Incorporation	O	P	P		Professional Narrative	
Part 2: Pattern Incorporation	O	P	P		Professional Narrative	
Part 3: Nature Interaction	O	–	P		Professional Narrative	
89, Adaptable Spaces						
Part 1: Stimuli Management	–	O	O	Architect		Spot Check
Part 2: Privacy	–	O	O	Architect		Spot Check
Part 3: Space Management	–	O	O	Owner		Spot Check
Part 4: Workplace Sleep Support	–	O	O	Owner		Spot Check
90, Healthy Sleep Policy						
Part 1: Non-Workplace Sleep Support	–	O	O		Policy Document	
91, Business Travel						
Part 1: Travel Policy	–	O	O		Policy Document	
92, Building Health Policy						
Part 1: Health Benefits	–	O	O		Policy Document	
93, Workplace Family Support						
Part 1: Parental Leave	–	O	O		Policy Document	
Part 2: Employer Supported Child Care					Policy Document	
Part 3: Family Support	–	O	O		Policy Document	
94, Self-Monitoring						
Part 1: Sensors and Wearables	–	O	O		Policy Document	
95, Stress and Addiction Treatment						
Part 1: Mind and Behavior Support	–	O	O		Policy Document	
Part 2: Stress Management	–	O	O		Policy Document	
96, Altruism						
Part 1: Charitable Activities	–	O	O		Policy Document	
Part 2: Charitable Contributions	–	O	O		Policy Document	

Figure11-1. (*continued on next page*)

...WELL Building Standard Features: Mind Concept and Innovation						
Features	Project Type			Verification Documentation		
	Core and Shell	New and Existing Interiors	New and Existing Buildings	Letter of Assurance	Annotated Documents	On-Site Assessment
97, Material Transparency						
Part 1: Material Information	O	O	O	Architect, Contractor, and Owner		
Part 2: Accessible Information	O	O	O			Visual Inspection
98, Organizational Transparency						
Part 1: Transparency Program Participation	–	O	O		Policy Document	Spot Check
99, Beauty and Design II						
Part 1: Ceiling Height	O	O	O		Architectural Drawing	Spot Check
Part 2: Artwork	O	O	O		Professional Narrative	
Part 3: Spatial Familiarity	O	O	O		Professional Narrative	
100, Biophilia II–Quantitative						
Part 1: Outdoor Biophilia	O	O	O	Owner		Spot Check
Part 2: Indoor Biophilia	–	O	O	Architect		Spot Check
Part 3: Water Feature	O	O	O	Architect		Spot Check
101–105, Innovation Features I–V						
Part 1: Innovation Proposal	O	O	O		Innovation Proposal	
Part 2: Innovation Support	O	O	O		Innovation Proposal	

Figure 11-1. The Mind Concept in WELL includes 17 features that promote strategies such as beauty through design, biophilia, healthy sleep, material and organizational transparency, and altruism. There are also five Innovation features that promote the evolution of the WELL Building Standard that project teams can pursue to achieve a higher certification level.

Part 2: Health and Wellness Library

Part 2: Health and Wellness Library of Feature 84 requires that a digital or physical library of books and magazines focusing on mental and physical health be provided to building occupants. At least one book title or magazine subscription is required for every 20 occupants, though no more than 20 titles are required.

BY THE NUMBERS

- *Life expectancy for individuals with mental illnesses is more than 10 years shorter than for those without mental health illnesses.*

FEATURE 85, INTEGRATIVE DESIGN

Feature 85, Integrative Design, is a precondition for all three project typologies that requires project stakeholders to meet during all phases of the building construction process. These meetings between stakeholders are called charrettes. A *charrette* is an intensive, multiparty workshop that brings people together to explore, generate, and collaboratively produce building design options. Charrettes are an essential part of the integrative design process.

The *integrative design process* is a comprehensive approach to the design, construction, and operation of building systems and equipment that encourages a holistic approach to maximizing building performance, human wellness and comfort, and environmental benefits. The integrative design process ideally starts before schematic design and continues through building design, construction, and operations and maintenance (post-construction). Starting this process in the early design phases allows project teams to clarify their priorities and goals and to identify relationships among building systems before designs are too far ahead.

Part 1: Stakeholder Charrette

Part 1: Stakeholder Charrette of Feature 85 requires that the stakeholders of a project meet for a charrette. At a minimum, the owner, architects, engineers, and facilities management team are required to attend. The stakeholder charrette must be held prior to the design and programming of the project.

Part 1 lists three tasks that the stakeholder charrette must accomplish. These three tasks include the following:

- Perform a values assessment and alignment exercise within the team to inform of any project goals as well as strategies to meet occupant expectations. This task helps the project team and stakeholders define what they consider to be important to the project.
- Discuss the needs of the occupants, focusing on wellness. This task is the starting point for defining wellness goals.
- Set future meetings to stay focused on the project goals and to engage future stakeholders who join the process after the initial meeting, such as contractors and subcontractors. This task continues the integrative design process through the scheduling of future charrettes at different stages of the project.

Part 2: Development Plan

Part 2: Development Plan of Feature 85 requires that a written document detailing the building's health-oriented mission be created. This document must be produced with the consent of all stakeholders. The plan should contain the following information as it pertains to the building's health mission:

- building site selection, including public transportation
- the seven WELL Concepts
- plans for implementing the project team's analyses and decisions
- operations and maintenance plans for facility managers and wellness-related building policy requirements

Part 3: Stakeholder Orientation

Parts 3: Stakeholder Orientation of Feature 85 requires that once building construction is complete, the project team tour the building as a group. During or after the tour, the project team must discuss how the building operations will support adherence to the building's WELL features.

FEATURE 86, POST-OCCUPANCY SURVEYS

Feature 86, Post-Occupancy Surveys, is a precondition for the New and Existing Interiors and New and Existing Buildings project types that requires the use of occupancy surveys to offer insight into the success of WELL features and provide feedback for improvements. The first part of this feature addresses the content of the survey. The second part addresses the reporting of the survey results.

Post-occupancy surveys are used to assess whether a building space has met the health and comfort needs of the occupants. The surveys also provide feedback to the project team and the International WELL Building Institute (IWBI) to help improve implementation of the WELL Building Standard. An additional benefit of the surveys is that they can have a positive impact on the building occupants' well-being and may increase productivity.

Part 1: Occupant Survey Content

Part 1: Occupant Survey Content of Feature 86 applies to buildings with 10 or more occupants. The survey used to fulfill the requirements of Part 1 is the Occupant Indoor Environmental Quality (IEQ) Survey™ from the Center for the Built Environment (CBE) at the University of California, Berkeley (though an approved alternate may be used instead). **See Figure 11-2.** The survey must be completed by at least 30% of the occupants at least once per year. However, thermal comfort should be surveyed twice per year. The survey includes the following topics:

- acoustics
- thermal comfort (including humidity and airflow)
- furnishings
- workspace light levels and quality
- odors, stuffiness, and other air quality concerns
- cleanliness
- layout

Part 2: Information Reporting

Part 2: Information Reporting of Feature 86 requires that the project team submit the results of the occupant surveys within 30 days to the building owners, building managers, and IWBI. If requested, the results must be reported to the building occupants as well. **See Figure 11-3.**

FEATURE 87, BEAUTY AND DESIGN I

Feature 87, Beauty and Design I, is a precondition for the New and Existing Interiors and New and Existing Buildings project types. This feature is derived from the Beauty and Spirit Imperative of the Living Building Challenge™, a building certification program and advocacy tool from the International Living Future Institute. To achieve this feature, a project must include design elements that thoughtfully create unique and culturally rich spaces.

OCCUPANT SURVEYS

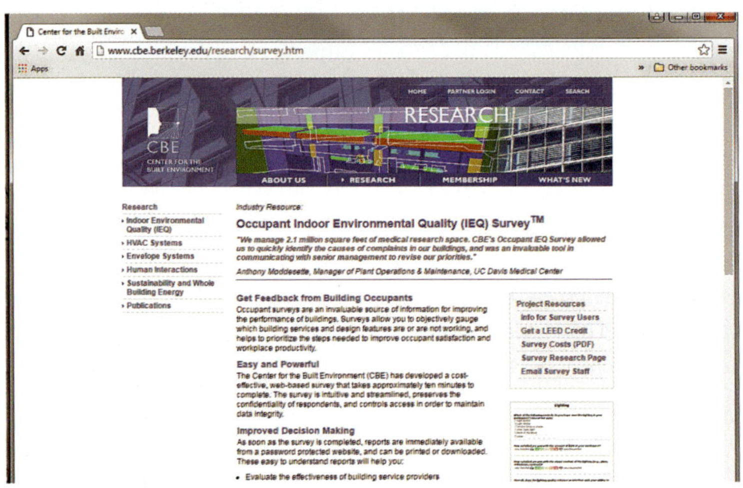

Figure 11-2. The survey that is used for Part 1: Occupant Survey Content of Feature 86, Post-Occupancy Surveys, is the Occupant Indoor Environmental Quality (IEQ) Survey™ from the CBE at UC Berkeley (or an approved alternate).

SURVEY INFORMATION REPORTING

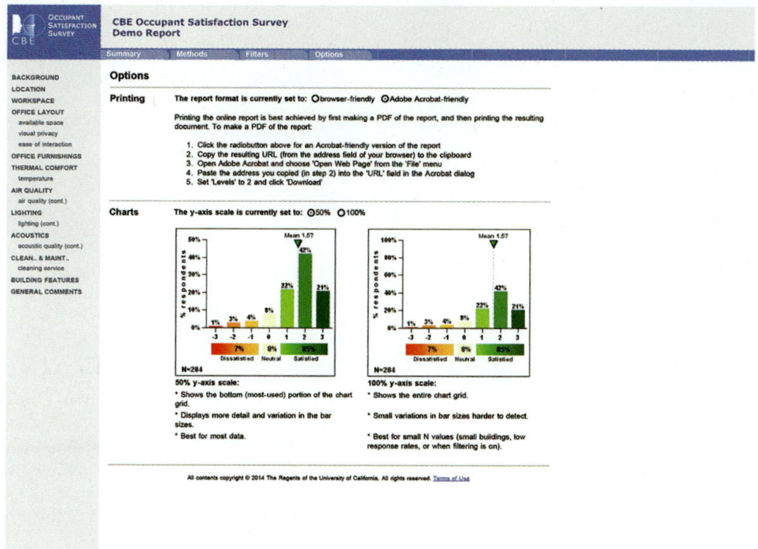

Figure 11-3. Building surveys must be reported to the building owners, building managers, and the IWBI. Upon request, the results must be reported to the building occupants as well.

Part 1: Beauty and Mindful Design

Part 1: Beauty and Mindful Design of Feature 87 requires that a narrative that describes the incorporation of thoughtful design elements meant to create a calm, peaceful environment be provided. These design elements must contain features intended for the following:

- human delight
- celebration of culture
- celebration of spirit
- celebration of place
- meaningful integration of public art

FEATURE 88, BIOPHILIA I—QUALITATIVE

Biophilia is human affinity for the natural world. Humans tend to desire and benefit from proximity and connection to life and lifelike processes, especially those found in the natural environment. Indoor environments that do not incorporate a sense of nature are perceived as cold and sterile, which diminishes mood and happiness. However, the use of biophilic design elements can help increase employee productivity, decrease stress, and reduce the amount of healing and recovery time in healthcare settings. **See Figure 11-4.**

Feature 88, Biophilia I—Qualitative, is an optimization for the Core and Shell project type, but it is a precondition for the New and Existing Interiors and New and Existing Buildings types. Core and Shell projects must meet all three parts of this feature to achieve it as an optimization. New and Existing Interiors are required to only meet the first two parts, but New and Existing Buildings must meet all three parts. To achieve this feature, a project team must develop a professional narrative outlining the biophilia plan.

BIOPHILIC DESIGN ELEMENTS

Figure 11-4. The use of biophilic design elements can help increase employee productivity, decrease stress, and reduce the amount of healing and recovery time in healthcare settings.

Part 1: Nature Incorporation

Part 1: Nature Incorporation of Feature 88 requires that the biophilia plan include a description of how the project will incorporate nature through the following elements:

- environmental elements
- lighting
- space layout

Part 2: Pattern Incorporation

Part 2: Pattern Incorporation of Feature 88 requires that the biophilia plan include a description of how the project will incorporate patterns of nature throughout the building design. Patterns of nature can be incorporated throughout the building design by using the principles of biomimicry. *Biomimicry* is the imitation of natural or biological designs, patterns, and processes in the production of materials or structures.

Part 3: Nature Interaction

Part 3: Nature Interaction of Feature 88 requires that the biophilia plan provide sufficient opportunities for human-nature interactions for the following areas:

- within the building
- within the project boundary, external to the building

FEATURE 89, ADAPTABLE SPACES

Feature 89, Adaptable Spaces, is an optimization for the New and Existing Interiors and New and Existing Buildings project types. This feature promotes building designs that allow for a variety of working spaces and address the minimum needs of occupants for privacy, storage, and rest. When building occupants are provided with a variety of workspace types, they can choose the best space to meet their needs at the time.

Part 1: Stimuli Management

Part 1: Stimuli Management of Feature 89 applies to regularly occupied spaces that are 186 m^2 (2000 ft^2) or larger. Part 1 requires that project teams provide documentation of the methods used to develop the following plans and establish the following zones:

- **Programming plans.** A programming plan must be developed using data collected from interviews, surveys, focus groups, and observations to establish workplace culture, work patterns, work processes, and space utilization.
- **Annotated floor plans.** Floor plans must be annotated with research data for establishing work zones for supporting a variety of work functions.
- **Designated quiet zones.** Quiet zones must be designated in enclosable or semi-enclosable rooms with no more than three seats per room.
- **Designated collaboration zones.** A collaboration zone is a space within a building that encourages group interplay and discussion through its strategic layout and design. Collaboration zones must be designated in enclosable or semi-enclosable rooms with at least three seats per room and at least one visual surface area for the communication of ideas or work. **See Figure 11-5.**

Part 2: Privacy

Part 2: Privacy of Feature 89 applies to projects with gross floor areas greater than 1860 m^2 (20,000 ft^2). Part 2 requires a designated quiet space for the purposes of focus, contemplation, and relaxation. This quiet space gives building occupants a place to rest and recharge during the workday. The quiet space must meet the following four requirements:

- The quiet space must be a minimum of 7 m^2 (75 ft^2) plus 0.1 m^2 (1 ft^2) per regular building occupant, not to exceed 74 m^2 (800 ft^2). For example, a quiet space for a building with 50 occupants must be 12 m^2 (125 ft^2) in area [$(50 \times 0.1) + 7 = 12$].
- The quiet space must have ambient lighting that provides continuously dimmable light levels at 2700 K or less.
- The noise criteria (NC) from mechanical systems in the quiet space must not exceed 30.
- A plan for the quiet space that includes a description of how the space incorporates two of three specific elements must be developed. The three specific elements are a plant wall and/or floor plantings, an audio device playing nature sounds, or a variety of seating arrangements.

COLLABORATION ZONES

Figure 11-5. Collaboration zones must be in enclosable or semi-enclosable rooms, contain at least three seats, and contain one visual vertical surface area.

Part 3: Space Management

Part 3: Space Management of Feature 89 requires that a project's work spaces have enough storage for employees in order to minimize clutter and maintain an organized work environment. One of two provisions must be used to meet this requirement. The first option is to provide a workstation cabinet with a minimum volume of 0.1 m^3 (4 ft^3) for each regular occupant. The second option is to provide each regular building occupant with a personal locker with a minimum volume of 0.1 m^3 (4 ft^3).

Part 4: Workplace Sleep Support

Part 4: Workplace Sleep Support of Feature 89 requires that employees be provided with furniture options that allow them to take short naps during the workday. Studies have shown that short naps improve mental and physical acuity and can lead to increased productivity, lower stress levels, and enhanced mood. At least one of the following furniture options must be provided for the first 30 occupants and an additional option must be provided for every 100 occupants thereafter:

- couch
- cushioned roll-out mat
- sleep pod
- fully reclining chair
- hammock

FEATURE 90, HEALTHY SLEEP POLICY

Feature 90, Healthy Sleep Policy, is an optimization for the New and Existing Interiors and New and Existing Buildings project types. Adults typically require 7 hours to 9 hours of sleep per night, depending on age. Adequate, quality sleep is important for maintaining mental and physical performance. *Sleep hygiene* is a set of personal habits and practices that help maximize sleep quality. Poor sleep hygiene can lead to a higher risk of weight gain, depression, heart disease, high blood pressure, and stroke. This feature addresses the need for sleep by limiting the amount of work that can be done at night and promotes monitoring behaviors that affect sleep patterns.

Part 1: Non-Workplace Sleep Support

Part 1: Non-Workplace Sleep Support of Feature 90 requires limits on the amount of work that can be performed at night and a subsidy for sleep-related behavior monitoring applications. Part 1a institutes an organizational cap at midnight for late-night work and communications. Part 1b requires that employers provide a 50% subsidy on software and/or applications that monitor daytime sleep-related behavior patterns.

FEATURE 91, BUSINESS TRAVEL

Feature 91, Business Travel, is an optimization for the New and Existing Interiors and New and Existing Buildings project types. Travelling for business, especially for extended periods, can cause increased stress for employees and their families. Business travel can also be physically stressful for an employee if there is limited access to healthy food or fitness facilities. The goal of this feature is to reduce the negative effects of business travel on employees.

Part 1: Travel Policy

Part 1: Travel Policy of Feature 91 addresses strategies and policies that reduce the stress of business travel. These strategies and policies include the following requirements:

- Traveling employees must be given the option of selecting non-red-eye flights or the option of working remotely on the day of arrival after a red-eye flight. Since red-eye flights are overnight flights, they can cause severe disruption to travelers' sleep cycles.
- The amount of travel time for short-term business trips must be limited so that it does not exceed both 5 hours and 25% of the total trip duration.
- Employees on long business trips must be given the opportunity to return home for

at least 48 hours once every two weeks for domestic travel and once every four weeks for international travel at the company's expense. Another option employees must be offered is for the company to fly out a friend or family member to meet the employee.

- Employees must be booked at hotels with free fitness centers or reimbursed for gym usage fees incurred during their travel.

FEATURE 92, BUILDING HEALTH POLICY

Feature 92, Building Health Policy, is an optimization for the New and Existing Interiors and New and Existing Buildings project types. Although many companies already have workplace health policies in place, this feature encourages a comprehensive health benefits plan that may better meet the needs of employees and their families.

BY THE NUMBERS

- *The chance that a person will be affected by a mood disorder, such as a major depressive disorder or bipolar disorder, is estimated at nearly 21% in the United States.*

Part 1: Health Benefits

Part 1: Health Benefits of Feature 92 requires that employers provide at least three of the following options:

- employer-based health insurance for part- and full-time workers, as well as their spouses and dependents, or subsidies to purchase individual insurance through an exchange
- flexible spending accounts or another benefit plan that is designed to reimburse employees for qualified medical expenses
- health savings accounts
- on-site immunizations or time off during the workday for immunizations
- workplace policies that encourage ill employees to stay home or work remotely

FEATURE 93, WORKPLACE FAMILY SUPPORT

Feature 93, Workplace Family Support, is an optimization for the New and Existing Interiors and New and Existing Buildings project types. Balancing work and home life can be difficult for some. This difficulty may increase stress levels and reduce productivity. A comprehensive plan for workplace family support can help employees properly care for themselves and their families. This feature addresses parental leave, childcare, and family support programs.

Part 1: Parental Leave

Part 1: Parental Leave of Feature 93 addresses short- and long-term paternity and maternity leave. While there are federal and state regulations concerning parental leave, they may include restrictions that limit which employees can access these benefits. For example, in the United States, the federal Family and Medical Leave Act (FMLA) only applies to businesses of 50 or more employees and only includes unpaid leave. Part 1 requires that 6 workweeks of paid paternity and maternity leave and an additional 12 workweeks of unpaid leave during any 12-month period be provided to all employees.

Part 2: Employer Supported Child Care

Part 2: Employer Supported Child Care of Feature 93 addresses the need for on-site child care or financial assistance for off-site child care. Part 2 requires that employers provide either an on-site licensed child care facility or financial assistance for child care in the form of subsidies or vouchers. Significant benefits can be realized when employers support child care. Since on-site child care reduces commuting times for parents, employee productivity increases. Financial assistance for off-site child care decreases the financial burden on the employees, which improves their stress levels and mental health.

Part 3: Family Support

Part 3: Family Support of Feature 93 addresses the need for family support. Part 3 expands the support provided by Part 1 to include other family members. Employers must provide the following:

- 12 workweeks of unpaid leave to care for a seriously ill child, spouse, domestic partner, parent, parent-in-law, grandparent, grandchild, or sibling
- the option to use paid sick time to care for the family members above listed
- breaks of at least 15 minutes every 3 hours for nursing mothers

FEATURE 94, SELF-MONITORING

Feature 94, Self-Monitoring, is an optimization for the New and Existing Interiors and New and Existing Buildings project types. Self-monitoring devices, such as wearable activity trackers, promote individuals' awareness of the biomarkers associated with their health and wellness. Individuals can monitor their health and physical activities through self-monitoring devices, thereby allowing them to make positive behavioral changes. **See Figure 11-6.**

SELF-MONITORING DEVICES

Figure 11-6. Self-monitoring devices enable individuals to monitor their own health and activity and to make positive behavioral changes.

Part 1: Sensors and Wearables

Part 1: Sensors and Wearables of Feature 94 requires that employers subsidize at least 50% of the cost of self-monitoring devices for each employee. These devices must contain sensors capable of measuring at least two of the following parameters:

- body weight/mass
- activity and steps
- heart rate
- sleep duration, quality, and regularity

FEATURE 95, STRESS AND ADDICTION TREATMENT

Feature 95, Stress and Addiction Treatment, is an optimization for the New and Existing Interiors and New and Existing Buildings project types. Chronic stress can have a significant impact on a person's health and wellness. Therefore, it is important that employers have programs in place to deal with stress and all of its negative manifestations, such as substance addiction. This feature provides a complement to other workplace health policies by addressing stress levels and addiction.

BY THE NUMBERS

- *Each year, about 14% of deaths, or 8 million lives lost, are attributed to mental health illnesses.*

Part 1: Mind and Behavior Support

Part 1: Mind and Behavior Support of Feature 95 requires that employers implement a comprehensive employee assistance program to assist employees in attaining professional help to manage chronic stress and other behavioral issues. An *employee assistance program (EAP)* is an employer-sponsored workplace program that is designed to assist employees in identifying and resolving personal issues that may be affecting mental and emotional well-being, which may affect their job performance. An EAP can help decrease employee absences, increase employee well-being and productivity, and reduce medical costs by addressing issues such as substance abuse, chronic stress, grief, family problems, and psychological disorders.

Part 2: Stress Management

Part 2: Stress Management of Feature 95 specifically addresses the need for overall stress management in the workplace. Part 2 requires that a stress management program, including access to a qualified counselor who offers group or private workshops and referrals, be provided to employees.

BY THE NUMBERS

- *Around the world in 2010, mental health illness and substance abuse cost nearly 184 million disability-adjusted life years (DALYs), 8.6 million years of life lost (YLL) to premature mortality, and over 175 million years lived with a disability (YLD).*

FEATURE 96, ALTRUISM

Feature 96, Altruism, is an optimization for the New and Existing Interiors and New and Existing Buildings project types.

Altruism is the selfless act of helping another person. Studies have shown that helping others has positive health and wellness benefits, in addition to promoting a sense of community and goodwill toward others. **See Figure 11-7.** Forms of altruism include volunteering as an individual or group with charitable organizations or making financial contributions. This feature encourages employers to institute policies that enhance community identity and promote social cohesion.

Part 1: Charitable Activities

Part 1: Charitable Activities of Feature 96 requires that employees be allowed to spend some time volunteering with charities while being paid by their company. Employers must provide the option of taking 8 hours of paid time off for participation in volunteer activities with registered organizations twice per year.

Part 2: Charitable Contributions

Part 2: Charitable Contributions of Feature 96 requires employers to commit to annually matching financial donations made to qualifying organizations by their employees. Many employers who institute a charitable contributions program set minimum and maximum amounts that they will match.

ALTRUISM

NREL

Figure 11-7. Helping others through activities such as volunteering for a charitable organization can have positive health and wellness benefits, in addition to promoting a sense of community and goodwill toward others.

FEATURE 97, MATERIAL TRANSPARENCY

Feature 97, Material Transparency, is an optimization for all three project types. The idea that building occupants have the right to know whether materials with harmful ingredients have been used to construct their indoor environment has recently become prevalent. Manufacturers can provide material information to product declaration programs for transparency. **See Figure 11-8.** Project teams can use this information to make decisions about the materials used in their buildings. This information must also be shared with all of the building occupants.

Part 1: Material Information

Part 1: Material Information of Feature 97 requires that at least 50%, by cost, of interior finishes and finish materials, furnishings, workstations, and built-in furniture uses one or more of the following product declaration options:

- **Declare® label.** A *Declare label* is a product declaration label designed by the International Living Future Institute that promotes transparency in disclosing material ingredients.
- **Health Product Declaration®.** A *Health Product Declaration (HPD)* is an open-standard format from the Health Product Declaration Collaborative for reporting product content and associated health information for building products and materials.
- **LEED v4 Materials and Resources (MR) Credit.** Any method accepted in USGBC's MR Credit—Building Product Disclosure and Optimization—Material Ingredients, Option 1: Material Ingredient Reporting, from the LEED v4 BD+C rating system can be used. **See Figure 11-9.**

In order to contribute to the feature, the product and its declaration must meet a minimum requirement. The product must indicate that all ingredients have been evaluated and disclosed down to 1000 ppm.

Part 2: Accessible Information

Part 2: Accessible Information of Feature 97 requires that all declaration information be compiled and made readily available to occupants either digitally or as part of a printed manual. Building occupants can access the digital or printed declaration information to relieve any stress or worries regarding any perceived environmental health hazards. This knowledge can help elevate building occupants' moods and well-being.

BY THE NUMBERS

- *Major depressive disorders affect nearly 16 million adults in the United States, causing sleep troubles, difficulty concentrating, loss of energy and appetite, low self-esteem, hopelessness, and physical pain.*

FEATURE 98, ORGANIZATIONAL TRANSPARENCY

Feature 98, Organizational Transparency, is an optimization for the New and Existing Interiors and New and Existing Buildings project types. *Organizational transparency* is the full disclosure of information and open communication by an employer with other stakeholders, such as employees, and the public. Employees with companies that have high levels of organizational transparency and positive social equity practices are more satisfied and have a greater sense of loyalty. These employees also have higher productivity and a lower risk of poor health.

Part 1: Transparency Program Participation

Part 1: Transparency Program Participation of Feature 98 requires that the organization seeking WELL certification participate in one of the two following transparency programs:

- **JUST™ Program.** The *JUST Program* is a voluntary transparency labeling platform for organizations created by the International Living Future Institute that includes information on employee treatment, safety, social stewardship, and diversity.

- **G4 Sustainability Reporting Guidelines.** The *G4 Sustainability Reporting Guidelines* is a reporting standard from the Global Reporting Initiative that helps organizations set goals, measure performance, and manage change in order to make their operations more sustainable for the environment, society, and the economy.

FEATURE 99, BEAUTY AND DESIGN II

Feature 99, Beauty and Design II, is an optimization for all three project types. This optional feature builds on the theme of pleasing visual design elements required by Feature 87, Beauty and Design I. Feature 99 addresses design elements, such as proportional ceiling height, artwork that enhances interior spaces, and distinctive or similar design elements, that aid in spatial familiarity.

DECLARE LABELS

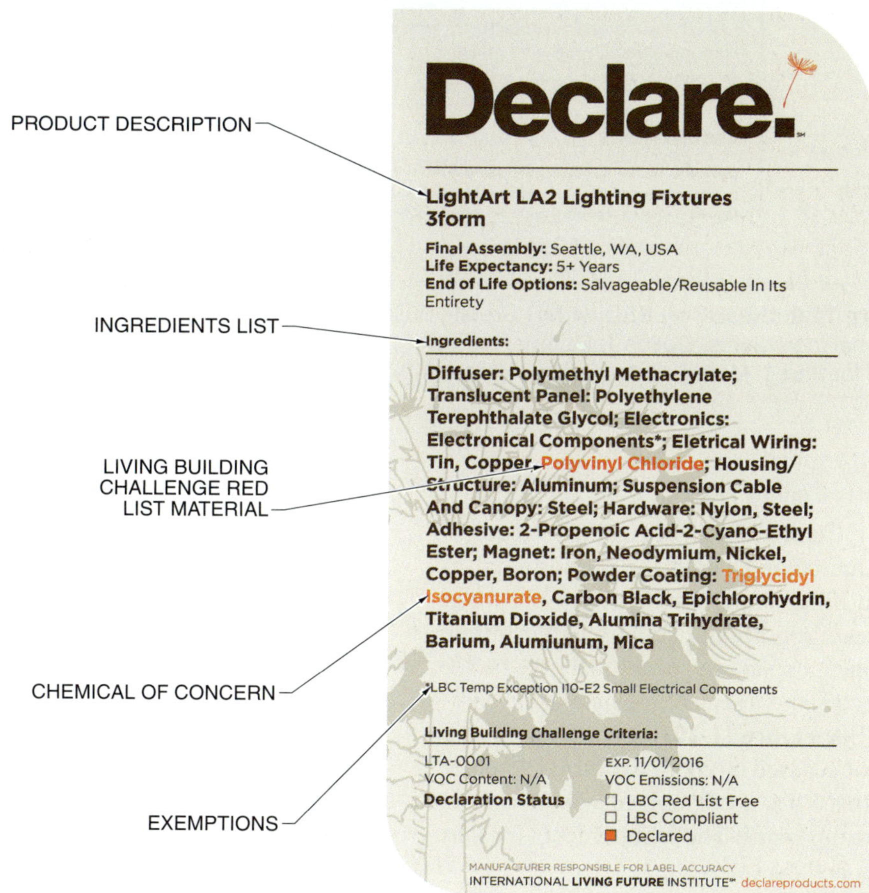

International Living Future Institute

Figure 11-8. Material transparency information may be found in manufacturer information provided through product declaration programs, such as the Declare label program from the International Living Future Institute.

MATERIAL INFORMATION

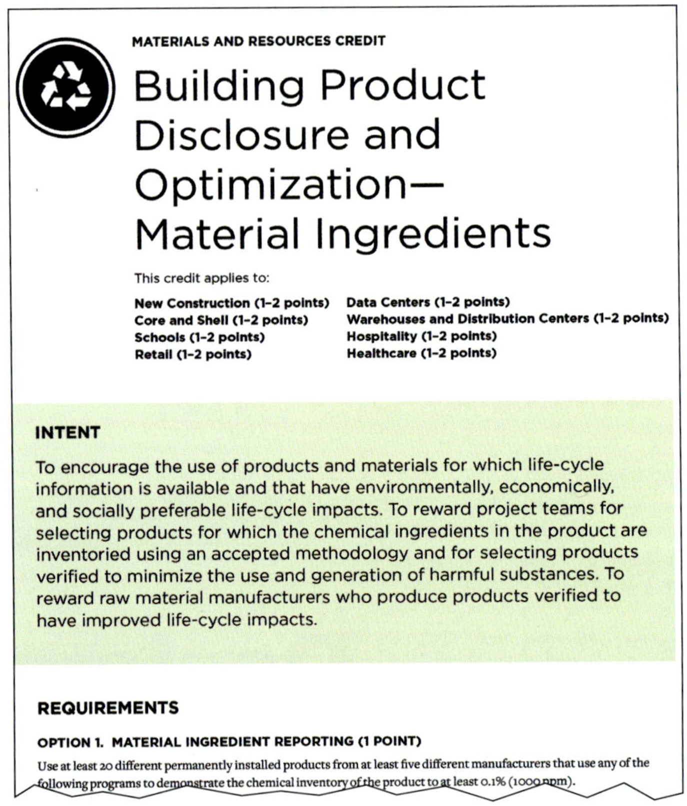
MATERIALS AND RESOURCES CREDIT

Building Product Disclosure and Optimization—Material Ingredients

This credit applies to:

New Construction (1-2 points)
Core and Shell (1-2 points)
Schools (1-2 points)
Retail (1-2 points)
Data Centers (1-2 points)
Warehouses and Distribution Centers (1-2 points)
Hospitality (1-2 points)
Healthcare (1-2 points)

INTENT

To encourage the use of products and materials for which life-cycle information is available and that have environmentally, economically, and socially preferable life-cycle impacts. To reward project teams for selecting products for which the chemical ingredients in the product are inventoried using an accepted methodology and for selecting products verified to minimize the use and generation of harmful substances. To reward raw material manufacturers who produce products verified to have improved life-cycle impacts.

REQUIREMENTS

OPTION 1. MATERIAL INGREDIENT REPORTING (1 POINT)

Use at least 20 different permanently installed products from at least five different manufacturers that use any of the following programs to demonstrate the chemical inventory of the product to at least 0.1% (1000 ppm).

U.S. Green Building Council

Figure 11-9. USGBC's LEED v4 MR Credit—Building Product Disclosure and Optimization–Material Ingredients, Option 1: Material Ingredient Reporting, lists several options that can be used for Part 1: Material Information of Feature 97, Material Transparency.

Part 1: Ceiling Height

Part 1: Ceiling Height of Feature 99 requires that the ceiling height be proportional to room dimensions for regularly occupied spaces. Proportional ceiling heights allow for open, expansive spaces that give building occupants a sense of openness and comfort. **See Figure 11-10.** Proportional ceiling height is based on room width and whether the space has a full wall of windows or is an atrium space. The requirements for Part 1 are as follows:

- The ceiling height for a room with a width of 9 m (30′) or less must be at least 2.7 m (8.8′).
- The ceiling height for a room with a width greater than 9 m (30′) must be at least 2.75 m (9′) plus 0.15 m (0.5′) for every additional 3 m (10′) of room width. For example, a room with a width of 18 m (60′) requires a ceiling height of 3.2 m (10.5′).
- The ceiling height for a room with a full-wall window view to the outdoors or an atrium space must be at least 2.75 m (9′) if the room has a width of 12 m (40′) or less. If the room has a width over 12 m (40′), the ceiling height must have an additional 0.15 m (0.5′) for every additional 4.5 m (15′) of room width. For example, a room with a width of 30 m (98.4′) with a full-wall window view of the outdoors requires a ceiling height of 3.35 m (11′).

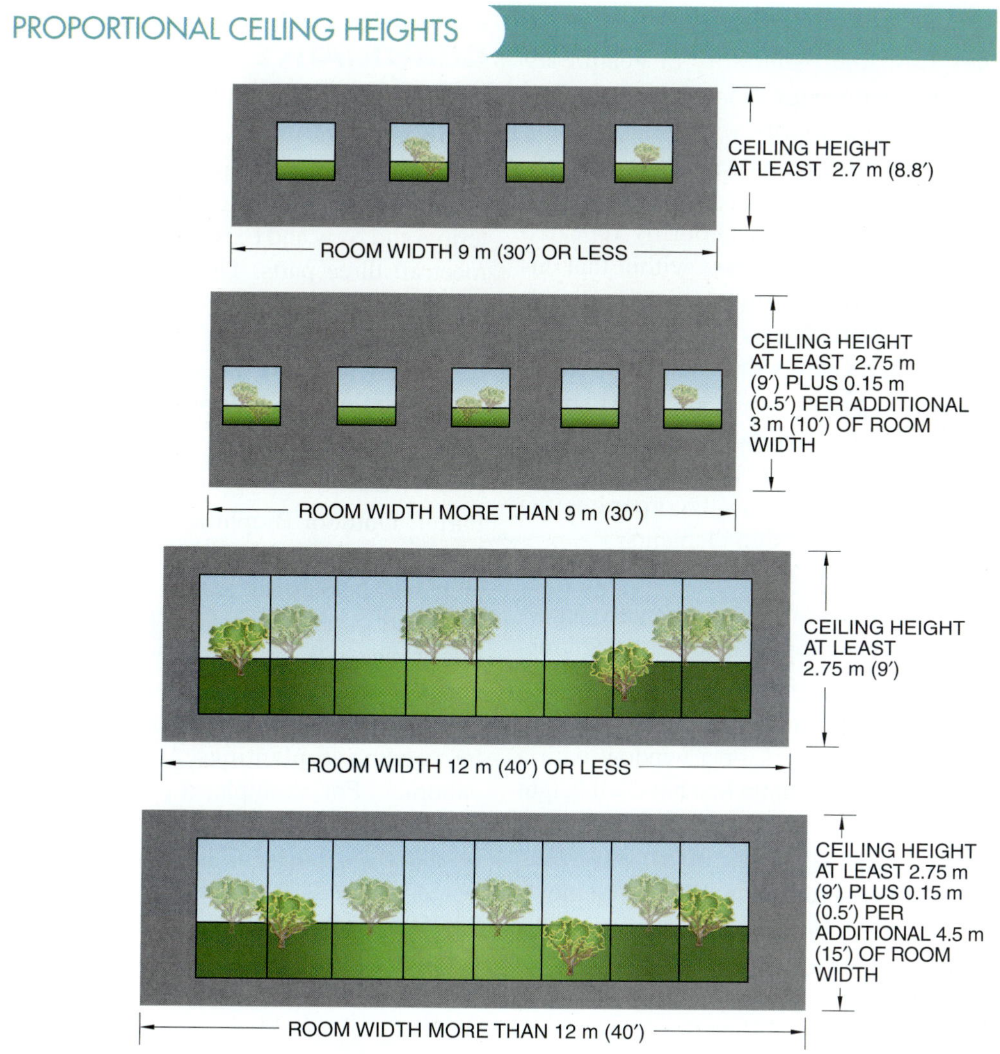

Figure 11-10. Part 1: Ceiling Height of Feature 99, Beauty and Design II, specifies room width to ceiling height proportions that allow for open, expansive spaces to give building occupants a sense of openness and comfort.

Part 2: Artwork

Part 2: Artwork of Feature 99 requires the integration of artwork throughout the common areas of a building, as well as other large regularly occupied spaces. Artwork adds visual interest to the interior space. Part 2 requires that a plan be developed to describe how artwork is incorporated in the following areas:

- entrances and lobbies
- all regularly occupied spaces greater than 28 m^2 (300 ft^2)

Artwork can be incorporated into building spaces for visual interest.

Part 3: Spatial Familiarity

Part 3: Spatial Familiarity of Feature 99 requires that design elements reinforcing the spatial familiarity within the building be included in the project. *Spatial familiarity* is a state of knowledge about an environment brought about by repeated association with elements within that environment. Maximizing spatial familiarity within large buildings helps building occupants and visitors find their way around the building without too much difficulty. Part 3 applies to buildings that are 929 m^2 (10,000 ft^2) or larger and requires that a plan be developed that incorporates the following three design elements:

- artwork that is distinctive
- zones or areas that use visually unifying design components such as lighting, furniture color, and flooring pattern/color
- corridors that are over 9 m (30′) in length ending in artwork or a window with a view to the outdoors that has a sill height no higher than 0.9 m (3′); windows must have a vista of at least 30 m (100′); long corridors without any visual elements can be confusing and disorienting to the building occupants

FEATURE 100, BIOPHILIA II—QUANTITATIVE

Feature 100, Biophilia II—Quantitative, is an optimization for all three project types. Core and Shell projects must meet Part 1 and Part 3 of this feature, while New and Existing Interiors and New and Existing Building must meet all three parts. This feature focuses on supporting occupants' emotional and psychological well-being through the use of outdoor biophilia, indoor biophilia, and water features. **See Figure 11-11.**

Part 1: Outdoor Biophilia

Part 1: Outdoor Biophilia of Feature 100 addresses the use of biophilic design elements in outdoor areas of the project site. Part 1 requires that at least 25% of the project site area features landscaped grounds or rooftop gardens. At least 70% of the 25% of the project site area must be vegetative plantings, including tree canopies. For example, a project with a project site area of 95,000 m^2 (312,000 ft^2) requires 23,750 m^2 (78,000 ft^2) of landscaped grounds or rooftop gardens. Of the 23,750 m^2 (78,000 ft^2), 16,625 m^2 (54,600 ft^2) is required to be vegetated.

Figure 11-11. Feature 100, Biophilia II—Quantitative, focuses on supporting occupants' emotional and psychological well-being through the use of outdoor biophilia, indoor biophilia, and water features.

Part 2: Indoor Biophilia

Part 2: Indoor Biophilia of Feature 100 addresses the use of biophilic design elements such as potted plants, planted beds, and plant walls in indoor spaces. Part 2a requires that potted plants or planted beds cover at least 1% of the floor area per floor. Part 2b requires that one plant wall per floor cover an area of wall equivalent to at least 2% of the floor area or the largest available wall, whichever is greater.

Part 3: Water Feature

Part 3: Water Feature of Feature 100 addresses the use of water biophilia in indoor spaces. Part 3 applies to projects larger than 9290 m^2 (100,000 ft^2) and requires one water feature for every 9290 m^2 (100,000 ft^2) of floor area. Each water feature must be at least 1.8 m (5.8′) in height or 4 m^2 (43 ft^2) in area. The water feature must have some type of water sanitation technology to prevent the formation of waterborne pathogens.

INNOVATION

In previous versions of the WELL Building Standard, the Innovation features were included as part of the Mind Concept. In the current version, the Innovation features have been separated into their own section. There are five Innovation features available for project teams to achieve. Each Innovation feature counts as an optimization for any of the three project types. These features supplement the certification process and allow project teams to achieve a higher certification level.

FEATURES 101–105, INNOVATION FEATURES I–V

Feature 101 through Feature 105 are optimizations for all three project types. These five features address topics that are not specifically covered in the WELL Building Standard by a feature but may still be applicable under the WELL Concepts. They provide project teams the opportunity to incorporate up to five innovations into their projects in an attempt to achieve a higher certification level, which helps the WELL Building Standard evolve. These features include two parts that address the proposal of each innovation and the necessary supporting documentation.

Part 1: Innovation Proposal

Part 1: Innovation Proposal requires that the proposed feature meet one of the following requirements:

- goes above and beyond the current requirements of an existing WELL feature
- relates to the concept of wellness in a novel way, is not already covered in the WELL Building Standard, and impacts the occupants or general public in a positive manner

Part 2: Innovation Support

Part 2: Innovation Support specifies the support material needed for the proposed feature. Part 2 requires that the proposed innovations be fully substantiated by scientific, medical, and industry research and explained in detail. These innovations must be consistent with all applicable laws, regulations, and leading building design and management practices.

KEY TERMS AND DEFINITIONS

altruism: The selfless act of helping another person.

biomimicry: The imitation of natural or biological designs, patterns, and processes in the production of materials or structures.

biophilia: Human affinity for the natural world.

KEY TERMS AND DEFINITIONS *(continued)*

charrette: An intensive, multiparty workshop that brings people together to explore, generate, and collaboratively produce building design options.

collaboration zone: A space within a building that encourages group interplay and discussion through its strategic layout and design.

Declare label: A product declaration label designed by the International Living Future Institute that promotes transparency in disclosing material ingredients.

employee assistance program (EAP): An employer-sponsored workplace program that is designed to assist employees in identifying and resolving personal issues that may be affecting mental and emotional well-being, which may affect their job performance.

G4 Sustainability Reporting Guidelines: A reporting standard from the Global Reporting Initiative that helps organizations set goals, measure performance, and manage change in order to make their operations more sustainable for the environment, society, and the economy.

health literacy: The ability for people to obtain, study, and understand the health information and available health services that are communicated to them through health literature and presented options.

Health Product Declaration (HPD): An open-standard format from the Health Product Declaration Collaborative for reporting product content and associated health information for building products and materials.

integrative design process: A comprehensive approach to the design, construction, and operation of building systems and equipment that encourages a holistic approach to maximizing building performance, human wellness and comfort, and environmental benefits.

JUST Program: A voluntary transparency labeling platform for organizations created by the International Living Future Institute that includes information on employee treatment, safety, social stewardship, and diversity.

organizational transparency: The full disclosure of information and open communication by an employer with other stakeholders, such as employees, and the public.

sleep hygiene: A set of personal habits and practices that help maximize sleep quality.

spatial familiarity: A state of knowledge about an environment brought about by repeated association with elements within that environment.

CHAPTER REVIEW

Completion

______________ **1.** Part 2: Health and Wellness Library of Feature 84, Health and Wellness Awareness, requires at least one book title or magazine subscription for every ___ occupants, with a maximum of 20 titles.

______________ **2.** A(n) ___ is an intensive, multiparty workshop that brings people together to explore, generate, and collaboratively produce building design options.

______________ **3.** ___ surveys are used to assess whether a building space has met the health and comfort needs of the occupants.

______________ **4.** Feature 87, ___, is derived from the Beauty and Spirit Imperative of the Living Building Challenge™.

______________ **5.** ___ is human affinity for the natural world.

CHAPTER REVIEW *(continued)*

_______________ **6.** Part 1: Stimuli Management of Feature 89, Adaptable Spaces, applies to regularly occupied spaces that are ___ or larger.

_______________ **7.** The quiet space required by Part 2: Privacy of Feature 89, Adaptable Spaces, must have ambient lighting that provides continuously dimmable light levels at ___ K or less.

_______________ **8.** The second requirement for Part 1: Non-Workplace Sleep Support of Feature 90, Healthy Sleep Policy, requires that employers provide a(n) ___% subsidy on software and/or applications that monitor daytime sleep-related behavior patterns.

_______________ **9.** Feature 93, ___, addresses parental leave, childcare, and family support programs.

_______________ **10.** Part 1: Sensors and Wearables of Feature 94 requires that employers subsidize at least ___% of the cost of self-monitoring devices for each employee.

_______________ **11.** A(n) ___ is an employer-sponsored workplace program that is designed to assist employees in identifying and resolving personal issues that may be affecting mental and emotional well-being, which may affect their job performance.

_______________ **12.** ___ is the full disclosure of information and open communication by an employer with other stakeholders, such as employees, and the public.

_______________ **13.** Part 3 of Feature 99, Beauty and Design II, applies to buildings that are ___ or larger.

_______________ **14.** Part 1: Outdoor Biophilia of Feature 100, Biophilia II—Quantitative, requires that at least ___% of the project site area features landscaped grounds or rooftop gardens.

Short Answer

1. List the positive impacts of a healthy state of mind.

2. List the increased health risks caused by stress and anxiety.

3. List the problems that mood disorders can cause within the workplace.

CHAPTER REVIEW *(continued)*

4. List the three tasks of the charrette required by Part 1: Stakeholder Charrette of Feature 85, Integrative Design.

5. List the information required in the written document for Part 2: Development Plan of Feature 85, Integrative Design.

6. List the information that should be included on an occupant survey.

7. Explain the design elements that should be considered for Feature 87, Beauty and Design I.

8. List the three parts of Feature 88, Biophilia I—Qualitative.

9. Explain the importance of sleep hygiene.

10. Sensors and other wearable self-monitoring devices required by Feature 94, Self-Monitoring, must measure at least two of which four parameters?

11. List the three product declaration options under Part 1: Material Information of Feature 97, Material Transparency.

12. Explain the benefits of organizational transparency and list the two options that are available under Part 1: Transparency Program Participation of Feature 98, Organizational Transparency.

CHAPTER REVIEW *(continued)*

13. List the requirements of Part 1: Ceiling Height of Feature 99, Beauty and Design II.

14. List and describe the three requirements of Feature 100, Biophilia II—Quantitative.

15. What are the proposal and support requirements that project innovations must meet for Features 101 through 105, Innovation Features I–V?

WELL AP EXAM PRACTICE QUESTIONS

1. Which feature is one of the preconditions for Core and Shell certification in the Mind Concept?
   A. Feature 84, Health and Wellness Awareness
   B. Feature 88, Biophilia—Qualitative
   C. Feature 94, Self-Monitoring
   D. Feature 98, Organizational Transparency

2. After sending out and receiving back post-occupant surveys, what is the next step for the project team?
   A. Report the results of the survey to the building occupants and local building officials within 30 days.
   B. Report the results of the survey to GBCI and the building owners within 45 days.
   C. Report the results of the survey to GBCI and the building owners within 30 days.
   D. Report the results of the survey to IWBI and the building owners within 30 days.

3. Which annotated document is required for the verification documentation of all three parts of Feature 88, Biophilia I–Qualitative?
   A. Professional narrative
   B. Policy document
   C. Architectural drawing
   D. Innovation proposal

4. Which feature of the Mind Concept creates a productive work environment that is free of distracting stimuli and includes spaces that are designed for focused work?
   A. Feature 87, Beauty and Design I
   B. Feature 89, Adaptable Spaces
   C. Feature 92, Building Health Policy
   D. Feature 100, Biophilia II—Quantitative

5. What type of materials and products are required to be disclosed under Part 1 of Feature 97, Material Transparency?
   A. Interior finishes, furnishings, and workstations
   B. Built-in furniture, interior finishes, and plumbing fixtures
   C. Interior finishes, furnishings, and electrical components
   D. Exterior brick, furnishings, and workstations

CHAPTER TWELVE

PRACTICE EXAM

1. What are three of the environmental pollutants that the WELL assessor measures during the performance test in order to satisfy the requirements of Feature 01, Air Quality Standards?

A. Formaldehyde, carbon monoxide, and carbon dioxide
B. Carbon monoxide, $PM_{2.5}$, and ozone
C. Formaldehyde, PM_{10}, and mold spores
D. VOCs, radon, and PFCs

2. Which feature employs the use of materials or cleaning procedures for high-touch surfaces that react to or disrupt microbes, thus killing them and preventing them from reproducing?

A. Feature 26, Advanced Material Safety
B. Feature 27, Antimicrobial Activity for Surfaces
C. Feature 28, Cleanable Environment
D. Feature 29, Cleaning Equipment

3. What is one of the transparency programs in which a project can participate to fulfill the requirements of Feature 98, Organizational Transparency?

A. JUST™ Program
B. Declare® label
C. GreenScreen®
D. National Organic Program (NOP)

4. Which precondition in the Comfort Concept requires that New and Existing Interiors and New and Existing Buildings projects develop an acoustic plan that identifies the loud and quiet zones within the project as well as any noise-producing equipment in the spaces?

A. Feature 74, Exterior Noise Intrusion
B. Feature 75, Internally Generated Noise
C. Feature 79, Sound Masking
D. Feature 81, Sound Barriers

5. How can visual interest be added to an entrance lobby for the requirements of Feature 99, Beauty and Design II?

A. Integrate a water feature.
B. Incorporate patterns of nature.
C. Integrate artwork.
D. Incorporate potted plants or planted beds.

6. How is each WELL feature achieved?

   A. Through documentation only
   B. Through performance verification only
   C. Through the WELL report completed by the WELL assessor
   D. Through documentation, performance verification, or a combination of both

7. Which three pedestrian amenities can be used to meet the requirements of Part 1: Pedestrian Amenities of Feature 67, Exterior Active Design?

   A. A bus stop, bike path, and drinking fountain
   B. A cluster of movable chairs and tables, bike paths, and drinking fountain
   C. A bench, bus stop, and public art
   D. A bench, cluster of movable tables and chairs, and drinking fountain

8. What is the verification method for the annotated document required by Feature 102, Innovation Feature II?

   A. Scientific paper
   B. Operations schedule
   C. Innovation proposal
   D. Architectural drawing

9. What is one of the requirements for the innovation proposal in Features 101–105, Innovation Features I–V?

   A. The proposed feature must relate to wellness in a novel way, must not already be covered in the WELL Building Standard, and must positively impact building occupants or the general public.
   B. The proposed feature must relate to the concept of environmental sustainability in a novel way and impact the surrounding site in a positive manner.
   C. The proposed feature must alter the requirements for a precondition so that it can be achieved as an optimization.
   D. The proposed feature must replace a precondition for the project type that is relevant to the project.

10. What does the acronym UVGI stand for?

    A. Ultraviolet gamma irradiation
    B. Ultraviolet germicidal irradiation
    C. Ultraviolet germicidal irritation
    D. Under value germicidal irradiation

11. What are two design elements that can be incorporated into a project's stairways and paths of frequent travel to meet the requirement of Part 3: Facilitative Aesthetics of Feature 64, Interior Fitness Circulation?

    A. Artwork and view windows
    B. Artwork and light levels of at least 200 lux
    C. View windows and wayfinding signage
    D. A water feature and artwork

12. A project team is attempting to achieve WELL certification for the New and Existing Interiors type. They have currently achieved all of the applicable preconditions and 60% of the applicable optimizations. What can they do to reach the Platinum certification level?

    A. Nothing. They have reached the Platinum certification level.
    B. They can achieve an additional 20% of optimizations to reach the Platinum certification level.
    C. They can achieve all of the optimizations for one WELL Concept to reach the Platinum certification level.
    D. They can achieve one additional optimization in each WELL Concept to reach the Platinum certification level.

13. The project team for a three-story office building is designing an entryway that will achieve Feature 08, Healthy Entrance. What strategies can they use to meet this goal?

A. They can use a permanent entryway system (with grilles) that is 3 m (10′) long and only half of the width of the entrance and a building entry vestibule with two normally closed doorways.
B. No entryway system is needed since the building will have an entry vestibule with two normally closed doorways.
C. They can use a permanent entryway system (with grilles) that is 3 m (10′) long and the width of the entrance and a building entry vestibule with two normally closed doorways.
D. They can use a permanent entryway system (with grilles) that is 2 m (8′) long and the width of the entrance and a revolving entrance door.

14. What is the purpose of a curative action plan?

A. To allow the project team to address unmet criteria after receiving the WELL report
B. To allow the project team to appeal the findings in the WELL report
C. To allow the project team to submit an idea for a new feature under a certain WELL Concept
D. To allow alternative solutions for meeting the intent of any feature requirement before registration

15. A pre-occupancy air flush must be performed under what conditions?

A. An outdoor temperature of at least 15°C (59°F) and relative humidity below 60% must be maintained for the entire flush.
B. An indoor temperature of at least 15°C (59°F) and relative humidity below 60% must be maintained for the entire flush.
C. An indoor temperature of at least 15°C (59°F) and relative humidity below 70% must be maintained for the entire flush.
D. An indoor temperature of at least 20°C (68°F) and relative humidity below 60% must be maintained for the first 12 hours of the flush.

16. Which feature in the Comfort Concept requires that the interior walls enclosing regularly occupied spaces reduce air gaps and limit sound transmission by properly sealing all acoustically rated partitions at the top and bottom tracks, staggering all gypsum board seams, and packing and sealing all penetrations through the wall?

A. Feature 74, Exterior Noise Intrusion
B. Feature 79, Sound Masking
C. Feature 80, Sound Reducing Surfaces
D. Feature 81, Sound Barriers

17. Feature 87, Beauty and Design I, is derived from which other green building rating system?

A. Leadership in Energy and Environmental Design (LEED)
B. Green Globes
C. Building Research Establishment Environmental Assessment Methodology (BREEAM)
D. Living Building Challenge

18. Which strategies are best suited to meet the requirements of Feature 17, Direct Source Ventilation?

A. Keeping copiers and printers in rooms equipped with ceiling fans
B. Keeping printers and copiers in isolated rooms with no ventilation
C. Keeping printers and copiers in separate rooms with a recirculated air system and self-closing door
D. Keeping copiers and printers in isolated rooms with direct venting and self-closing doors

19. Devices that measure indoor environmental conditions for Feature 18, Air Quality Monitoring and Feedback, must measure and record which three pollutant metrics at least once per hour?

A. Carbon monoxide, temperature, and formaldehyde
B. Particle count/mass, carbon dioxide, and ozone
C. Particle count/mass, carbon dioxide, and carbon monoxide
D. Ozone, formaldehyde, and carbon monoxide

20. Which three strategies can be used to achieve Feature 23, Advanced Air Purification?

A. Displacement ventilation systems, UVGI systems, and photocatalytic oxidation systems
B. Activated carbon filters, UVGI systems, and photocatalytic oxidation systems
C. MERV 8 filters, UVGI systems, and photocatalytic oxidation systems
D. Activated carbon filters, UVGI systems, and DOAS systems

21. Which combustion-based appliances are banned in regularly occupied spaces by Feature 24, Combustion Minimization?

A. Furnaces, stoves, space-heaters, ranges, and ovens
B. Fireplaces, boilers, space-heaters, ranges, and water heaters
C. Fireplaces, stoves, water heaters, ranges, and ovens
D. Fireplaces, stoves, space-heaters, ranges, and ovens

22. How many employees in a 100-employee office building must be able to find seating at any given time in an eating space that meets the requirements of Feature 52, Mindful Eating?

A. 25
B. 40
C. 50
D. 75

23. A project team for a New and Existing Interiors project aims to achieve Feature 26, Enhanced Material Safety. If $50,000 worth of furnishings, built-in furniture, interior features, and finish materials were included in the project, how much of this cost must meet one of the voluntary material reporting programs listed in the requirements for Feature 26?

A. $5,000
B. $12,500
C. $25,000
D. $50,000

24. When must the HVAC system be balanced in order to achieve Feature 03, Ventilation Effectiveness?

A. During the installation of the HVAC system
B. Before substantial completion of the project
C. After occupancy but before WELL certification of the project
D. After substantial completion but before occupancy of the project

25. Under which standard is building envelope commissioning of a commercial office project performed after substantial completion?

A. ASHRAE Guideline 0-2005
B. ASHRAE Standard 62.1-2013
C. ASHRAE Standard 55-2013
D. South Coast Air Quality Management District (SCAQMD) Rule 1168

26. What is the recommended approximate daily water consumption for men and women according to the Institute of Medicine (IOM)?

A. Women 1.7 L (57 oz); Men 2.7 L (125 oz)
B. Women 2.7 L (91 oz); Men 3.7 L (125 oz)
C. Women 4.7 L (159 oz); Men 5.7 L (193 oz)
D. Women 6.7 L (227 oz); Men 7.7 L (260 oz)

27. What two contaminants are measured in a water sample for Feature 30, Fundamental Water Quality?

A. Sediment and dissolved metals
B. Agricultural contaminants and microorganisms
C. Sediment and microorganisms
D. Sediment and disinfectants

28. Per the WELL Building Standard, exposure to dissolved metals leads to potential health effects in which body systems?

A. Cardiovascular, digestive, nervous, reproductive, and urinary
B. Cardiovascular, immune, integumentary, and urinary
C. Nervous, endocrine, reproductive, and skeletal
D. Integumentary, muscular, and skeletal

29. In order to meet the requirements of Feature 15, Increased Ventilation, the outdoor air supply rate must be increased over the rate required in Feature 03, Ventilation Effectiveness by what percentage?

A. 10%
B. 20%
C. 30%
D. 40%

30. Feature 25, Toxic Material Reduction, limits the amount of halogenated flame retardants in flooring, piping, and thermal insulation to what level?

A. 0.001% (10 ppm)
B. 0.01% (100 ppm)
C. 0.1% (1000 ppm)
D. 1.0% (10000 ppm)

31. A project's water sample is tested by the WELL assessor. The sample indicates that lead levels are 0.005 mg/L and copper levels are less than 0.9 mg/L. What type of filter can be used to best meet the requirements of Feature 31, Inorganic Contaminants?

A. KDF filters
B. RO filtration system
C. GAC filters
D. No filter strategy is needed since stated levels are within required limits.

32. What by-product is formed when chlorine reacts with organic matter in a water supply?

A. Chloramine
B. Fluoride
C. Trihalomethane
D. 2,4-dichlorophenoxyacetic acid

33. What are the methods of documentation and performance verification for the two parts of Feature 52, Mindful Eating?

A. An architectural drawing, a letter of assurance from the contractor, and an on-site spot check
B. An operations schedule and an on-site visual inspection
C. An architectural drawing, a letter of assurance from the owner, and an on-site spot check
D. An on-site spot check and a letter of assurance from the architect

34. How often does Feature 35, Periodic Water Quality Testing, require that water testing take place?

A. Monthly, with reports submitted to IWBI quarterly
B. Monthly, with reports submitted to IWBI annually
C. Quarterly, with reports submitted to IWBI annually
D. Quarterly, with reports submitted to IWBI every three years

35. What water condition suggests that the water may contain other types of dangerous pathogens such as bacteria, viruses, and protozoa?

A. High salinity
B. High turbidity
C. Low pH
D. Presence of coliforms

36. Where in the building does Feature 37, Drinking Water Promotion, require the project team to locate drinking water dispensers?

A. Within 15 m (50′) of every elevator
B. Within 30 m (100′) of all parts of regularly occupied floor space
C. Within 30 m (100′) of all parts of unoccupied floor space
D. Within 75 m (250′) of all parts of regularly occupied floor space

37. What methods of fruit and vegetable promotion must be included in the design for an owner-operated cafeteria in an office building project?

A. Visible and accessible salad bar, at least one prominent display of vegetables, and fruits placed at the front of the foodservice line
B. Visible and accessible salad bar, nutritional labeling for meals incorporating vegetables, and fruits placed near checkout locations
C. Visible and accessible salad bar, prominent visual displays of fruits and vegetables, vegetables placed at the front of the foodservice line, and fruits placed at checkout locations
D. Visible and accessible salad bar, prominent visual displays of fruits and vegetables, nutritional labeling for raw fruits and vegetables, and large serving sizes for meals incorporating vegetables

38. A project team wishing to pursue Feature 51, Food Production, specifies an on-site garden that is 20 m^2 (215 ft^2). How many building occupants is the project team estimating for the project?

A. 20
B. 80
C. 120
D. 200

39. What type of substances are polychlorinated biphenyls, styrene, and benzene?

A. Inorganic contaminants
B. Organic contaminants
C. Pesticides
D. Disinfectants

40. The foodservice for an institutional building project offers 16 main course options on its menu. How many low-calorie versions must be offered for the project to fulfill the requirements of Part 1: Meal Sizes of Feature 47, Serving Sizes?

A. 4
B. 8
C. 12
D. 16

41. Which precondition in the Nourishment Concept specifies that raw meat, fish, and poultry be separated from prepared foods in properly labeled cold storage spaces?

A. Feature 40, Food Allergies
B. Feature 42, Food Contamination
C. Feature 46, Safe Food Preparation Materials
D. Feature 50, Food Storage

42. Which feature in the Nourishment Concept requires a label for a food containing monosodium glutamate (MSG) and sodium nitrate?

A. Feature 43, Artificial Ingredients
B. Feature 45, Food Advertising
C. Feature 48, Special Diets
D. Feature 52, Mindful Eating

43. Which project team member is responsible for authorizing the registration of a project seeking WELL certification?

A. Project administrator
B. WELL AP
C. Architect
D. Owner

44. What must be included on the nutrition information label for food and beverage items to fulfill the requirements of Feature 44, Nutritional Information?

A. Total calories, no trans fat, total sugar content, and any artificial ingredients
B. Macronutrients, total sugar content, and any artificial ingredients
C. Total calories, macronutrients, and total sugar content
D. No trans fat, allergen information, vitamin content, and artificial ingredients

45. The levels of which two substances are included in the quarterly testing and reporting for Feature 35, Periodic Water Quality Testing?

A. Lead and arsenic
B. Mercury and benzene
C. Copper and fluoride
D. Copper and benzene

46. How many deaths worldwide does the World Health Organization (WHO) attribute to insufficient fruit and vegetable intake?

A. 1.1 million
B. 2.7 million
C. 5.2 million
D. 8.9 million

47. What combination of cooking tool material and cutting board material may be specified to fulfill the requirements of Feature 46, Safe Food Preparation Materials?

A. An uncoated aluminum cooking tool and a marble cutting board
B. An uncoated copper cooking tool and a nonstick steel cutting board
C. A cast iron cooking tool and a glass cutting board
D. A nonstick stainless steel cooking tool and a coated plastic cutting board

48. Which feature is one of the preconditions for all three project types in the Nourishment Concept?

A. Feature 38, Fruits and Vegetables
B. Feature 40, Food Allergies
C. Feature 45, Food Advertising
D. Feature 48, Special Diets

49. What is the minimum CRI R9 value that meets the requirements of Feature 58, Color Quality?

A. 25
B. 35
C. 50
D. 80

50. Which water treatment strategy is best used for the removal of organic chemicals?

A. UVGI system
B. GAC filter
C. MERV 13 filter
D. KDF filters

51. How many grams (g) of sugar at the most may be in individually sold, single-serving, non-beverage food items per Part 1: Refined Ingredient Restrictions of Feature 39, Processed Foods?

A. 30
B. 33
C. 40
D. 55

52. The sale of foods and beverages that contain trans fat is prohibited in the cafeteria of a New and Existing Buildings project. Which methods will the WELL assessor use to document and verify this trans fat ban?

A. A letter of assurance from the owner and an architectural drawing
B. A letter of assurance from the owner and an on-site visual inspection
C. An operations schedule and an on-site spot check
D. An operations schedule and an on-site performance test

53. What type of documents are design drawings, policy documents, and operations schedules?

A. Letters of assurance
B. General documents
C. Annotated documents
D. Registration documents

54. The ambient light at the workstations in an office building project is measured at 280 lux (26 fc). What amount of lux (fc) must be supplied by task lighting at the workstations for this project to fulfill the third requirement of Part 1: Visual Acuity for Focus for Feature 53, Visual Lighting Design?

A. 300 lux to 500 lux (28 fc to 46 fc)
B. 400 lux to 500 lux (37 fc to 46 fc)
C. 500 lux to 600 lux (46 fc to 56 fc)
D. None. This project already meets the requirements with 280 lux (26 fc).

55. Which feature in the Light Concept requires that all computer screens on desks located within 4.5 m (15′) of view windows be able to be oriented within a 20° angle perpendicular to the plane of the nearest window and that overhead luminaires not be aimed directly at computer screens?

A. Feature 53, Visual Lighting Design
B. Feature 55, Electric Light Glare Control
C. Feature 57, Low-Glare Workstation Design
D. Feature 61, Right to Light

56. Which feature in the Fitness Concept utilizes financial reimbursements or subsidies to promote an increase in occupant physical activity?

A. Feature 65, Activity Incentive Programs
B. Feature 66, Structured Fitness Opportunities
C. Feature 67, Exterior Active Design
D. Feature 68, Physical Activity Spaces

57. Which part and feature is the only precondition for Core and Shell projects in the Light Concept?

A. Part 2: Glare Minimization of Feature 55, Electric Light Glare Control
B. Part 2: Daylight Management of Feature 56, Solar Glare Control
C. Part 2: Responsive Light Control of Feature 60, Automated Shading and Dimming Controls
D. Part 3: Uniform Color Transmittance of Feature 63, Daylight Fenestration

58. What is the term for the internal clock that keeps the human body's hormones and bodily processes on a roughly 24-hour cycle, even in continuous darkness?

A. Metabolic system
B. Circadian rhythm
C. Sleep-wake cycle
D. Photoreceptive cells

59. Intrinsically photosensitive retinal ganglion cells (ipRGCs) are sensitive to what color of light?

A. Violet-blue
B. Green-yellow
C. Teal-blue
D. Pinkish-red

60. What is the name of photoreceptive cells in a person's eyes that are mainly responsible for the synchronization of circadian rhythms?

A. Rods
B. Intrinsically photosensitive retinal ganglion cells (ipRGCs)
C. Photosensitive optic nerve cells
D. Cones

61. According to Part 1: Melanopic Light Intensity for Work Areas of Feature 54, Circadian Lighting Design, between which hours must the required light level be present every day of the year?

A. 6:00 AM to 11:00 AM
B. 8:00 AM to 12:00 PM
C. 9:00 AM to 1:00 PM
D. 10:00 AM to 3:00 PM

62. Which feature in the Fitness Concept requires access to on-site professional fitness programs and fitness education?

A. Feature 65, Activity Incentive Programs
B. Feature 66, Structured Fitness Opportunities
C. Feature 67, Exterior Active Design
D. Feature 68, Physical Activity Spaces

63. What is a strategy for reducing the amount of glare caused by interior electric light sources with luminance of 20,000 cd/m^2 to 50,000 cd/m^2?

A. Providing automated dimming controls
B. Installing interior window shading or blinds
C. Shielding the lamps at an angle of 5° or greater
D. Shielding the lamps at an angle of 15° or greater

64. Which part of Feature 56, Solar Glare Control, applies to glazing that is less than 2.1 m (7′) above the floor?

A. Part 1: Glare Avoidance
B. Part 1: Healthy Sunlight Exposure
C. Part 1: View Window Shading
D. Part 2: Glare Minimization

65. Which precondition in the Nourishment Concept requires labels for foods that contain peanuts, fish, shellfish, soy, milk and dairy products, eggs, wheat, tree nuts, or gluten?

A. Feature 39, Process Foods
B. Feature 40, Food Allergies
C. Feature 43, Artificial Ingredients
D. Feature 44, Nutritional Information

66. In order to maintain visual acuity, task lighting must provide 300 lux to 500 lux (28 fc to 46 fc) at the work surface when ambient light is below what light level?

A. 215 lux (20 fc)
B. 250 lux (23 fc)
C. 300 lux (28 fc)
D. 350 lux (33 fc)

67. What percentage of a building compared to its total lot size is the threshold for the first two parts of Feature 67, Exterior Active Design?

A. 25%
B. 40%
C. 75%
D. 90%

68. What is the minimum light reflectance value (LRV) required for vertical surfaces that are visible from regularly occupied spaces for Feature 59, Surface Design?

A. LRV of 0.7 for at least 50% of surface area
B. LRV of 0.7 for at least 70% of surface area
C. LRV of 0.8 for at least 80% of surface area
D. LRV of 0.9 for at least 50% of surface area

69. What percentage of people worldwide do not get the recommended 30-minutes per day minimum of moderate-intensity physical activity?

A. 20%
B. 40%
C. 60%
D. 80%

70. Part 2: Glare Minimization of Feature 55, Electric Light Glare Control, requires that bare lamps more than 53° above the center of view have luminance less than what level?

A. 4500 cd/m^2
B. 6500 cd/m^2
C. 8000 cd/m^2
D. 10,000 cd/m^2

71. Which humidity control strategy that meets the requirements of Feature 16, Humidity Control, can reduce dryness and irritation of the skin, prohibit the growth of microbial pathogens, and limit the amount of off-gassing from building products that contain formaldehyde?

A. Provide humidification when relative humidity is below 30% and dehumidification when relative humidity is over 50%.
B. Maintain a level of relative humidity of 65% year-round.
C. Provide humidification when relative humidity is above 30% and dehumidification when relative humidity is below 50%.
D. Provide humidification when relative humidity is above 50%. No dehumidification is needed under the feature.

72. People who engage in 2.5 hours of moderate-intensity physical activity per week can reduce their overall mortality risk by what percentage?

A. 10%
B. 20%
C. 30%
D. 40%

73. An office building project that is 6500 m^2 (70,000 ft^2) aims to fulfill the requirements of Feature 62, Daylight Modeling. How many square meters (square feet) of the project can be used to achieve $ASE_{1000,250}$ and still meet the requirements?

A. 163 m^2 (1750 ft^2)
B. 325 m^2 (3500 ft^2)
C. 650 m^2 (7000 ft^2)
D. 1300 m^2 (14,000 ft^2)

74. What is the minimum CRI Ra value (CRI average of reference colors R1 through R8) that meets the requirements of Feature 58, Color Quality?

A. 25
B. 35
C. 50
D. 80

75. Which feature in the Fitness Concept is an optimization that requires a dedicated indoor exercise space and access to external exercise spaces?

A. Feature 66, Structured Fitness Opportunities
B. Feature 67, Exterior Active Design
C. Feature 68, Physical Activity Spaces
D. Feature 69, Active Transportation Support

76. The occupants of an office building project pursuing New and Existing Interiors certification complain of musculoskeletal issues such as low back pain and sore shoulder muscles from sitting at their workstations in chairs that are not adjustable. Which precondition in the Comfort Concept benefits the wellness of these occupants by specifically addressing the issue of seat flexibility?

A. Feature 71, Active Furnishings
B. Feature 73, Ergonomics: Visual and Physical
C. Feature 85, Integrative Design
D. Feature 89, Adaptable Spaces

77. Feature 69, Active Transportation Support, requires bicycle storage and support on a project site or within how far of the building's main entrance?

A. 50 m (165′)
B. 100 m (325′)
C. 200 m (650′)
D. 300 m (950′)

78. Which common type of health issue results from poor ergonomics?

A. Asthma
B. Musculoskeletal disorders
C. Heart disease
D. Eye strain

79. How many preconditions features are required in the Comfort Concept for Core and Shell certification?

A. 2
B. 3
C. 4
D. 5

80. A project team must submit a letter of assurance from which professional for the documentation of Feature 72, Accessible Design?

A. MEP engineer
B. Contractor
C. Building inspector
D. Architect

81. To meet the requirements for Feature 74, Exterior Noise Intrusion, a project team must ensure that the average sound pressure level from outside noise intrusion does not exceed how many A-weighted decibels (dBA)?

A. 30
B. 40
C. 50
D. 60

82. Feature 64, Interior Fitness Circulation, applies to projects with how many floors?

A. 1 to 3
B. 1 to 4
C. 2 to 4
D. 3 to 6

83. Which feature in the Fitness Concept is an optimization that requires showers and bicycle storage on-site or near the building entrance?

A. Feature 66, Structured Fitness Opportunities
B. Feature 67, Exterior Active Design
C. Feature 68, Physical Activity Spaces
D. Feature 69, Active Transportation Support

84. What is required for the performance verification and/or documentation of Feature 77, Olfactory Comfort?

A. An architectural drawing
B. A letter of assurance from the architect
C. A policy document and a performance test
D. A visual inspection

85. What are the source separation methods that can be used to fulfill the requirements of Part 1: Source Separation of Feature 77, Olfactory Comfort?

A. Negative pressurization, acoustical ceiling tiles, ceiling baffles, vestibules, and hallways
B. Interstitial rooms, vestibules, thermal comfort devices, hallways, and self-closing doors
C. Negative pressurization, interstitial rooms, vestibules, hallways, and self-closing doors
D. Vestibules, self-closing doors, gaskets, sweeps, and non-hollow-core door slabs

86. What is the maximum allowable reverberation time (RT60) for conference rooms that still meets the requirement of Feature 78, Reverberation Time?

A. 0.4 seconds
B. 0.5 seconds
C. 0.6 seconds
D. 0.7 seconds

87. Which WELL Concept aims to maintain a healthy, distraction-free, and productive work environment by ensuring that the acoustic, ergonomic, olfactory, and thermal conditions of a building meet the needs of the building occupants?

A. Air
B. Fitness
C. Comfort
D. Mind

88. Which feature in the Mind Concept will be achieved if design elements intended for human delight; celebration of culture, spirit, and place; and meaningful integration of public art are incorporated into a project?

A. Feature 85, Integrative Design
B. Feature 87, Beauty and Design I
C. Feature 96, Altruism
D. Feature 99, Beauty and Design II

89. The biophilia plan for Part 1: Nature Incorporation of Feature 88, Biophilia I–Qualitative, requires descriptions for which three design elements?

A. Environmental elements, lighting, and quiet zones
B. Lighting, space layout, and building size
C. Environmental elements, lighting, and space layout
D. Space layout, building size, and quiet zones

90. What is the maximum noise criteria (NC) for enclosed offices that still meets the requirements of Part 2: Mechanical Equipment Sound Levels of Feature 75, Internally Generated Noise?

A. 35
B. 40
C. 45
D. 50

91. A project team is attempting to meet the requirements for Part 2: Privacy of Feature 89, Adaptable Spaces, for an office building that will have 80 regular building occupants. How many square meters (square feet) must be designated for a quiet space?

A. 7 m^2 (75 ft^2)
B. 12 m^2 (125 ft^2)
C. 15 m^2 (161 ft^2)
D. 74 m^2 (800 ft^2)

92. What are the two requirements for Feature 90, Healthy Sleep Policy?

A. Midnight cap for late-night work and communications; sleep pods for the first 30 regular building occupants
B. A 50% subsidy on software and/or applications that monitor daytime sleep-related behavior patterns; an acoustic plan that identifies both loud and quiet zones
C. Fully reclining chairs at every occupied workstation; midnight cap for late-night work and communications
D. Midnight cap for late-night work and communications; a 50% subsidy on software and/or applications that monitor daytime sleep-related behavior patterns

93. Which WELL project type applies to a building where 10% of the total floor area is occupied by a different tenant than the other 90% of total floor area?

A. Core and Shell
B. New and Existing Interiors
C. New and Existing Buildings
D. WELL Communities pilot

94. What are names of the three parts for Feature 93, Workplace Family Support, that intend to help employees maintain a positive work-life balance?

A. Parental Leave, Stress Management, and Charitable Activities
B. Employer Supported Child Care, Family Support, and Spatial Familiarity
C. Parental Leave, Employer Supported Child Care, and Family Support
D. Family Support, Non-Workplace Sleep Support, and Stress Management

95. An employer wishes to offer the use of self-monitoring fitness devices that track body weight/mass, heart rate, and activity and steps to all of its employees. Each of the devices costs $100. What is the minimum amount that the employer must subsidize per device to fulfill the requirements of Feature 94, Self-Monitoring?

A. $50
B. $75
C. $90
D. $100

96. Feature 95, Stress and Addiction Treatment, is an optimization for which project types?

A. Core and Shell and New and Existing Buildings
B. New and Existing Interiors and New and Existing Buildings
C. Core and Shell and New and Existing Interiors
D. All three project types

97. A project was registered for Core and Shell certification on January 15, 2017. Four years and three months have passed, and the project has not yet been scheduled for performance verification. What is the project team's option to still achieve project certification?

A. Request an extension from IWBI.
B. Schedule a site visit within the next nine months or the registration will expire.
C. Do nothing, since it is past the length of time to request performance verification.
D. Wait another year and then request performance verification.

98. What documentation and/or performance verification is required by Part 1: Material Information of Feature 97, Material Transparency?

A. Visual inspection
B. Architectural drawing
C. Policy document
D. Architect, contractor, and owner letters of assurance

99. A project team receives the project's WELL report. In the report, the WELL assessor did not give the project credit for the outdoor biophilia that was integrated into the building design. How can the project team attempt to resolve this situation?

A. Submit an appeal within 180 days of receiving the WELL report.
B. Submit an alternative adherence path within 180 days of receiving the WELL report.
C. Submit an innovation request within 90 days of receiving the WELL report.
D. Submit an appeal within 90 days of receiving the WELL report.

100. What must a project do in order to maintain certification for the WELL Building Standard?

A. Conduct yearly recertification performance verifications.
B. Register for recertification within 36 months (three years) of receiving initial certification.
C. Register for recertification within 42 months (three and a half years) of receiving initial certification.
D. Nothing, since the certification lasts forever.

APPENDIX

By The Numbers...

The By the Numbers items found in *WELL AP Exam Preparation Guide* reiterate specific facts and statistics about human health and wellness that are found in the WELL Building Standard. These facts and statistics may be encountered on the WELL AP exam. See Appendix H: Concept and Feature References in the WELL Building Standard for a complete list of works cited.

Air

- Air pollution contributes to about 50,000 premature deaths in the United States and approximately 7 million premature deaths worldwide (or one in eight premature deaths), making it the largest global environmental health risk.
- A person breathes on average more than 15,000 L of air every day.
- Even though smoking tobacco causes at least 400,000 premature deaths per year in the United States alone, at least 42 million U.S. adults (more than a billion worldwide) are smokers.
- A person who is a tobacco smoker has a life expectancy that is 10 years less than a nonsmoker.
- When cigarettes are smoked, their 600 ingredients form 7000 compounds that include at least 69 known carcinogens (cancer-causing substances).
- The levels of volatile organic compounds (VOCs) in an indoor environment can be five times higher than outdoor levels.
- U.S. agricultural and commercial industries use about 1 billion pounds of pesticide every year.
- Humidity control can be important for indoor spaces because the off-gassing of formaldehyde can be 1.8 to 2.6 times higher when relative humidity increases by 35%.
- Up to 60% of asthmatic people who live in cities or other urban environments also have reactions to cockroach allergens.
- The affinity for carbon monoxide to bind to the hemoglobin of red blood cells is 210 times stronger than that of oxygen, meaning it is 210 times more likely to be carried through the bloodstream than oxygen.
- Carbon monoxide from nonvehicle sources such as fuel-burning appliances and engine-powered equipment is responsible for an estimated 170 deaths per year in the United States.

Water

- The Institute of Medicine (IOM) recommends that women consume approximately 2.7 L (91 oz) of water per day and that men consume 3.7 L (125 oz) of water per day from all sources. Active people or people in warmer environments may require larger amounts of water.
- The World Health Organization (WHO) reports that almost one billion people lack access to safe drinking water worldwide and that two million annual deaths can be attributed to unsafe water, sanitation, and hygiene.
- While there are environmental implications to the overreliance on bottled water, another concern is that the quality of bottled water is subject to degradation. In one study, levels of antimony in 48 brands of bottled water from 11 European countries increased by 90% after 6 months of storage due to antimony leaching from polyethylene terephthalate (PET) bottles, which are designated as recyclable "1."
- In the 1990s, a U.S. Geological Survey detected pesticide compounds in virtually every stream in agricultural, urban, and mixed-use areas, as well as in 30% to 60% of the groundwater.

Nourishment

- Sugar often accounts for more than 500 calories in a person's daily diet, contributing greatly to the global obesity pandemic.
- For a quarter of the U.S. population, sugar-sweetened beverages (SSBs) such as fruit-flavored drinks, sodas, sports drinks, and energy drinks account for 200 calories consumed daily.
- The average person in the United States consumed nearly 2600 calories per day in 2010 versus about 2100 calories per day in 1970.
- According to calculations using the body mass index (BMI), more than two-thirds of adults in the United States are overweight, and more than a third are obese.
- Obesity is considered a global pandemic since more than 1.9 billion adults worldwide were considered overweight in 2014, and nearly 600 million of those adults were obese.
- The World Health Organization (WHO) reports that 2.7 million deaths worldwide are attributed to insufficient fruit and vegetable intake, which is why several features in the Nourishment Concept focus on the availability and consumption of fruits and vegetables.
- In the United States, only 8% of people consume the recommended 4 servings of fruit per day and only 6% consume the recommended 5 servings of vegetables per day.
- Even though the recommended limit for sugar intake is only 6 to 9 teaspoons per day, the average consumption in the United States is more than 22 teaspoons per day.
- Every year, approximately 30,000 people visit emergency rooms, 2000 are hospitalized, and 150 die due to allergic reactions to food.
- In the United States alone, foodborne illnesses account for 48 million people getting sick with 128,000 hospitalizations and 3000 deaths.

...By The Numbers

Light

- The Institute of Medicine (Health and Medicine Division), reports that about 50 to 70 million adults in the United States suffer from a chronic sleep or wakefulness disorder.
- Typically, an ambient light level of 300 lux is sufficient for most tasks.

Fitness

- It is recommended that all healthy adults engage in at least 30 minutes of moderate-intensity aerobic activity 5 days per week and muscle-strengthening activities at least 2 days per week.
- In the United States alone, fewer than 50% of elementary school students, 10% of adolescents, and 5% of adults obtain 30 minutes of daily physical activity.
- An average adult obtains only 6 to 10 minutes of moderate to vigorous intensity physical activity a day.
- Over 60% of all people worldwide do not get the recommended daily 30-minute minimum of moderate-intensity physical activity.
- Physical inactivity is an independent risk factor for numerous chronic diseases and is estimated to be responsible for 30% of ischemic heart disease (heart problems caused by narrowed arteries), 27% of type 2 diabetes, and 21% to 25% of breast and colon cancer cases.
- Lack of physical activity can increase the odds of having a stroke by 20% to 30% and shave off three to five years of life.
- Physical inactivity is the fourth leading risk factor for mortality, accounting for 6% to 9% of deaths worldwide, or 3 to 5 million mortalities every year.
- Just 2.5 hours of moderate-intensity physical activity per week can reduce overall mortality risk by nearly 20%.
- Sitting burns 50 fewer calories per hour than standing, and sitting for more than 3 hours per day is associated with a 2-year lower life expectancy.

Comfort

- Low back pain, one of the most common musculoskeletal disorders, affects about 31 million Americans at any given time.
- In 2013, nearly 380,600 days of work were missed in the United States (one-third of all days away from work) due to musculoskeletal disorders.
- In 2010, nearly 7% (more than 169 million) of all disability-adjusted life years (DALYs) worldwide resulted from musculoskeletal disorders.
- In 2006, only 11% of the office buildings surveyed in the United States provided thermal environments that met generally accepted goals of occupant satisfaction.
- The human body's core temperature has a narrow range of 36°C to 38°C (97°F to 100°F), which is regulated by a portion of the brain called the hypothalamus.

Mind

- Around the world in 2010, mental health illness and substance abuse cost nearly 184 million disability-adjusted life years (DALYs), 8.6 million years of life lost (YLL) to premature mortality, and over 175 million years lived with a disability (YLD).
- Life expectancy for individuals with a mental illness is more than 10 years shorter than for those without a mental health illness.
- Each year, about 14% of deaths, or 8 million lives lost, are attributed to mental health illnesses.
- The chance that a person will be affected by a mood disorder, such as a major depressive disorder or bipolar disorder, is estimated at nearly 21% in the United States.
- Major depressive disorders affect nearly 16 million adults in the United States, causing sleep troubles, difficulty concentrating, loss of energy and appetite, low self-esteem, hopelessness, and physical pain.

The WELL Building Standard Appendices

The Appendices in the WELL Building Standard provide supplementary information and references. This information can be helpful in understanding and applying WELL, but it is unnecessary for an exam candidate to attempt to memorize the entirety of the Appendices for the WELL AP exam.

Appendix A: Glossary

Appendix A contains a list of definitions for general terms, substances, and units and measures that are used in WELL or may be used on a project site. Exam candidates should be familiar with these terms.

Appendix B: Standards Citations

Appendix B contains a comprehensive list of citations for the requirements of WELL Concept features that are based on third-party standards.

Appendix C: Tables

Appendix C includes tables that provide more details or explanations for the requirements of WELL Concept features. Some tables that are particularly useful include a list of high-touch surfaces, a cleaning protocol, and explanations and charts of melanopic ratios.

Appendix D: Feature Types and Verification Methods

Appendix D is a helpful list of the methods of performance verification—letters of assurance, annotated documents, and on-site checks—for each feature in the WELL Concepts. Performance verification is the important step in the WELL certification process used to ensure that the strategies and design interventions applied to a project are performing as intended.

Appendix E: (Removed in Q1 2018 Addenda)

Appendix E was removed, but it still refers users to "Applying LEED and the WELL Building Standard™," a crosswalk document available at wellcertified.com/resources. As of Q1 2018, the remaining appendices have not been renamed.

Appendix F: (Removed in Q1 2018 Addenda)

Appendix F was removed, but it still refers users to "Living Building Challenge® & the WELL Building Standard™," a crosswalk document available at wellcertified.com/resources. As of Q1 2018, the remaining appendices have not been renamed.

Appendix G: External Reviewers

Appendix G is a list of critical reviewers from the sciences, building industry, and medical institutions who have contributed their expertise to the development of the WELL Building Standard.

Appendix H: Concept and Feature References

Appendix H contains a comprehensive list of the sources used for statistical or factual knowledge found throughout the WELL Concept backgrounds and many of the feature descriptions.

Appendix I: Core and Shell Scope

Appendix I clarifies the scope of each WELL feature for the Core and Shell type as they apply to a project. This helps project teams determine whether the requirements of a feature apply to the entire building, the extent of the developer buildout, the extent of the developer buildout with the capacity for tenant achievement, the extent of the developer buildout and confirmed tenant support, or common areas and spaces under owner control.

ANSWER KEY

CHAPTER 2—THE WELL BUILDING STANDARD

Chapter Review

Completion ______ 20

1. seven
2. 105
3. part
4. Innovation features
5. precondition
6. optimization
7. synergy
8. Trade-offs
9. 90
10. New and Existing Interiors
11. 25
12. six

Short Answer ______ 21

1. *Answers are as follows:*
   - Air Concept—29 features
   - Water Concept—8 features
   - Nourishment Concept—15 features
   - Light Concept—11 features
   - Fitness Concept—8 features
   - Comfort Concept—12 features
   - Mind Concept—17 features
   - Innovation—5 features
2. One of the features in the Air Concept specifies that a percentage of the furniture in a project meet voluntary material reporting requirements, such as those in Imperative 10 of the Living Building Challenge 3.0. Another feature in the Air Concept has the same requirements for permanent entry walk-off systems as those found in one of the Indoor Environmental Quality (EQ) credits for the LEED v4 Building Design and Construction (BD+C) rating system.
3. When adjustable workstation chairs are specified to achieve ergonomic requirements for the Comfort Concept, chairs made from non-VOC-emitting materials can be specified to also increase the indoor air quality for the Air Concept.
4. *Answers are as follows:*
   - New and Existing Buildings—41 preconditions and 59 optimizations
   - New and Existing Interiors—36 preconditions and 62 optimizations
   - Core and Shell—26 preconditions and 28 optimizations
5. *Answers are as follows:*
   - Multifamily Residential
   - Educational Facilities
   - Retail
   - Restaurant
   - Commercial Kitchens
   - Communities
6. The features of the WELL Concepts can be used to protect, as well as lessen the negative effects of the built environment on, human body systems such as the cardiovascular, digestive, endocrine, immune, integumentary, muscular, nervous, reproductive, respiratory, skeletal, and urinary systems.

WELL AP Exam Practice Questions ______ 22

1. A
2. C
3. D
4. D
5. B

CHAPTER 3—THE WELL CERTIFICATION PROCESS

Chapter Review

Completion ______ 32

1. stakeholder
2. Silver
3. charrette
4. project administrator
5. owner
6. WELL assessor
7. WELL AP
8. letter of assurance
9. Annotated documents
10. performance verification
11. alternative adherence path (AAP)
12. 180
13. curative action plan
14. appeal
15. three
16. WELL Community Standard™
17. Design Verfied

Short Answer ______ 33

1. *Answers are as follows:*
   - financial bottom line
   - environmental bottom line
   - social bottom line
2. A project that has successfully gone through the WELL certification process may be perceived as having added value. Buildings that have met rigorous green building standards, such as the WELL Building Standard, can be used for marketing and branding opportunities. Since the WELL Building Standard focuses on the health and wellness of the people working in and around a project, there may also be increases in productivity and employee retention. These increases can lead to higher profits. In addition, healthier employees can reduce the health insurance costs for the company and the employees.
3. Projects seeking certification for all project types can be certified to the Silver, Gold, or Platinum levels. A project is WELL Certified™ Silver when it meets all of the applicable preconditions but no optimizations. A project is WELL Certified™ Gold when it meets all of the applicable preconditions and 40% or more of applicable optimizations. A project is WELL Certified™ Platinum when it achieves all of the applicable preconditions and 80% or more of applicable optimizations.
4. *Answers are as follows:*
   - Step 1. Registration
   - Step 2. Documentation requirements
   - Step 3. Performance verification
   - Step 4. Certification
   - Step 5. Recertification and documentation submission
5. *Answers are as follows:*
   - architects
   - contractors
   - interior designers
   - safety/environmental compliance officers
   - wellness coordinators
   - acoustical consultants
   - mechanical, electrical, and plumbing (MEP) engineers

6. A WELL assessor provides technical support to a project team seeking WELL certification. A WELL assessor also ensures that a project meets the requirements of the WELL Building Standard by reviewing the project's documentation and visiting the project site to verify that the project performs as intended. Once adherence to the WELL Building Standard is verified, the WELL assessor creates the WELL report and determines the level of certification achieved.
7. The WELL report is a comprehensive report of a project that includes a feature-by-feature summary indicating whether the project team successfully provided documentation to verify the achievement of each feature or whether the project has successfully met the measurable criteria for specific features.
8. Features 65, 66, 90, 91, 92, 93, 94, 95, 96, and 98 are categorized as WELL organizational protocol.

WELL AP Exam Practice Questions __________ 34

1. B
2. C
3. A
4. C
5. B

CHAPTER 4—AIR—PRECONDITIONS

Chapter Review

Completion __________ **54**

1. 12
2. 7.5 m (25′)
3. 62.1-2013
4. 25
5. Minimum efficiency reporting value (MERV)
6. International WELL Building Institute (IWBI)
7. mold
8. Construction Pollution Management
9. 3 m (10′)
10. high-touch surface
11. 3
12. 100
13. three
14. fluorescent

Short Answer __________ **55**

1. *Answers are as follows:*
   - formaldehyde—less than 27 ppb
   - VOCs—less than 500 μg/m³
   - CO—less than 9 ppm
   - $PM_{2.5}$—less than 15 μg/m³
   - PM_{10}—less than 50 μg/m³
   - ozone—less than 51 ppb
   - radon—less than 0.148 Bq/L (4 pCi/L) at lowest occupied level
2. *Answers are as follows:*
   - Part 1: Interior Paints and Coatings
   - Part 2: Interior Adhesives and Sealants
   - Part 3: Flooring
   - Part 4: Insulation
   - Part 5: Furniture and Furnishings
3. The cooling coils in a building's HVAC system must be inspected for mold growth on a quarterly basis. Also the ceilings, walls, and floors of a building must be inspected for discoloration, mold, and signs of water damage or pooling.
4. *Answers are as follows:*
   - The three strategies for Part 1: Entryway Walk-off Systems of Feature 08, Healthy Entrance, include the following:
     - permanent entryway system comprising grilles, grates, or slots
     - rollout mats
     - material manufactured as an entryway walk-off system
   - The three strategies for Part 2: Entryway Air Seal of Feature 08, Healthy Entrances, include the following:
     - building entry vestibule with two normally closed doorways
     - revolving entrance doors
     - at least three normally shut doors that separate occupied spaces from the outdoors
5. *Answers are as follows:*
   - the Cleaning Equipment and Training section of Table A4 in Appendix C of the WELL Building Standard
   - a list of approved product seals with which all cleaning, disinfection, and hand-hygiene products must comply in accordance with the Cleaning, Disinfection and Hand Hygiene Product section in Table A4
   - a list of high-touch surfaces and a schedule of sanitization or disinfection as specified in the Disinfection and Sanitization section in Table A4
   - a cleaning schedule that specifies the extent and frequency of cleaning, including the Entryway Maintenance section of Table A4
   - dated, maintained cleaning logs that are available to all occupants
6. *Answers are as follows:*
   - EPA 40 CFR Part 745.65 is used as the basis for on-site investigations of lead in buildings.
   - EPA 40 CFR Part 745.227 is used as the basis for work practices and lead abatement activities.
7. *Answers are as follows:*
   - Thermometers, switches, or electrical relays that contain mercury are not specified or used.
   - Lamps are installed that comply with the low-mercury limits from Table A5 in Appendix C of the WELL Building Standard. A plan to replace any noncompliant lamps with low-mercury or mercury-free lamps must be developed.
   - Only illuminated exit signs that use light-emitting diode (LED) or light-emitting capacitor (LEC) lamps are specified.
   - Mercury vapor or probe-start, metal halide high-intensity discharge (HID) lamps are not used.
8. *Answers are as follows:*
   - site drainage and irrigation
   - local water table
   - building penetrations, such as windows, and mechanical, electrical, and plumbing (MEP) penetrations
   - porous building materials connected to exterior sources of water
9. *Answers are as follows:*
   - plumbing leaks
   - appliances, such as clothes washers, directly connected to the water supply
   - porous building materials connected to interior sources of water
   - new building materials that have high "built-in" moisture content or have been wetted during construction and brought into the interior
10. *Answers are as follows:*
    - high interior relative humidity levels
    - air leakage that could wet exposed interior materials or hidden interstitial materials through condensation
    - cooler surfaces such as basement or slab-on-grade floors or closets/cabinets on exterior walls
    - oversized AC units that cycle on and off too quickly, which prevents moisture from being removed from the interior air
11. *Answers are as follows:*
    - exposed entryways and glazing
    - porous cladding materials
    - finished floors in potentially damp or wet rooms
    - interior sheathing in potentially damp or wet rooms
    - sealing and storing of absorptive materials during construction

WELL AP Exam Practice Questions ______ 57

1. D
2. C
3. C
4. B
5. D

CHAPTER 5—AIR—OPTIMIZATIONS

Chapter Review

Completion ______ 75

1. 29
2. 60
3. Air Infiltration Management
4. 30
5. Pollution Isolation and Exhaust
6. three
7. all three
8. 100
9. 55-2013
10. 10
11. 30
12. interior finishes
13. 25
14. Antimicrobial Activity for Surfaces
15. 1.0
16. ammonia

Short Answer ______ 76

1. When humidity is too low, building occupants may experience dryness and irritation of the eyes, throat, and mucous membranes. When humidity is too high, it allows for the accumulation and growth of microbial pathogens such as bacteria and mold. Excessive humidity can also lead to increased off-gassing of building materials.
2. *Answers are as follows:*
   - particle count (35,000 counts per m^3 [1000 counts per ft^3] or finer) or particle mass (10 mg/m^3 or finer)
   - carbon dioxide (25 ppm or finer)
   - ozone (10 ppb or finer)
3. *Answers are as follows:*
   - ozone—51 ppb
   - PM_{10}—50 $\mu g/m^3$
   - temperature—±8°C (15°F) from set indoor building temperature
   - humidity—60% relative humidity
4. A qualified and registered professional mechanical engineer must review a dedicated outdoor air system (DOAS). The review must address thermal comfort, ventilation rates, and the system serviceability and reliability, as well as ensure the DOAS's compliance with the ventilation standards used in Feature 03, Ventilation Effectiveness.
5. *Answers are as follows:*
   - Perishable foods not in a refrigerator must be stored in sealed containers.
   - Indoor garbage cans less than 113 L (30 gal.) must have lids that can be operated without the use of hands or be enclosed by cabinetry in an undercounter, pull-out drawer that has a handle separate from the garbage can.
   - Indoor garbage cans greater than 113 L (30 gal.) must have a lid.
6. One method is to install activated carbon filters or combination particulate/carbon filters in ductwork as a part of the main HVAC system. The other method is to install properly sized, standalone air purifiers with carbon filters in all regularly occupied spaces.
7. *Answers are as follows:*
   - fireplaces
   - stoves
   - space-heaters
   - ranges
   - ovens
8. *Answers are as follows:*
   - internal combustion engines
   - furnaces
   - boilers, steam generators, and process heaters
   - water heaters
9. *Answers are as follows:*
   - Part 1: Perfluorinated Compound Limitation
   - Part 2: Flame Retardant Limitation
   - Part 3: Phthalate (Plasticizers) Limitation
   - Part 5: Urea-Formaldehyde Restriction
10. *Answers are as follows:*
    - One of three labels from the International Living Future Institute, either a Declare: Living Building Challenge Red List Free label; a Declare: Living Building Challenge Compliant label; or a Living Product Challenge label
    - Cradle to Cradle™ Material Health Certified with a V2 Gold or Platinum or V3 Bronze, Silver, Gold, or Platinum Material Health Score
    - No GreenScreen® Benchmark 1, List Translator 1, or List Translator Possible 1 substances over 1000 ppm, as verified by a qualified professional
11. The high-touch surfaces in a building that must be coated with or consist of abrasion-resistant, nonleaching materials that meet EPA requirements for antimicrobial activity include all countertops and fixtures in bathrooms and kitchens, handles, doorknobs, light switches, and elevator buttons.
12. *Answers are as follows:*
    - smooth and free of visible defects that can make the surface difficult to clean
    - smooth welds and seams to allow for easy cleaning
    - no sharp internal angles, corners, or crevices that can trap dirt or pathogens

WELL AP Exam Practice Questions ______ 78

1. C
2. A
3. D
4. C
5. D

CHAPTER 6—WATER

Chapter Review ______ 95

Completion

1. Turbidity
2. higher
3. 1.0
4. total coliforms
5. pathogen
6. kinetic degradation fluxion (KDF)
7. organic
8. Granular activated carbon (GAC)
9. agricultural
10. fertilizer
11. 50
12. disinfectant
13. disinfectant by-product (DBP)
14. Fluoride
15. 4.0
16. inorganic metal

17. activated carbon filter
18. suspended solids
19. three
20. Legionella
21. 500
22. one

Short Answer ________________________________ **96**

1. *Answers are as follows:*
   - Feature 30, Fundamental Water Quality
   - Feature 31, Inorganic Contaminants
   - Feature 32, Organic Contaminants
   - Feature 33, Agricultural Contaminants
   - Feature 34, Public Water Additives
2. *Answers are as follows:*
   - Feature 35, Periodic Water Quality Testing
   - Feature 36, Water Treatment
   - Feature 37, Drinking Water Promotion
3. *Answers are as follows:*
   - lead—less than 0.01 mg/L
   - arsenic—less than 0.01 mg/L
   - antimony—less than 0.006 mg/L
   - mercury—less than 0.002 mg/L
   - nickel—less than 0.012 mg/L
   - copper—less than 1.0 mg/L
4. *Answers are as follows:*
   - styrene—less than 0.0005 mg/L
   - benzene—less than 0.001 mg/L
   - ethylbenzene—less than 0.3 mg/L
   - polychlorinated biphenyls (PCBs) —less than 0.0005 mg/L
   - vinyl chloride—less than 0.002 mg/L
   - toluene—less than 0.15 mg/L
   - xylenes (total: m, p, and o)—less than 0.5 mg/L
   - tetrachloroethylene—less than 0.005 mg/L
5. *Answers are as follows:*
   - atrazine—less than 0.001 mg/L
   - simazine—less than 0.002 mg/L
   - glyphosate—less than 0.70 mg/L
   - 2,4-dichlorophenoxyacetic acid—less than 0.07 mg/L
6. Chlorine and chloramine are the two disinfectants listed in Part 1: Disinfectants of Feature 34, Public Water Additives. Chlorine must be less than 0.06 mg/L; chloramine must be less than 4.0 mg/L. While chloramine is not as effective at killing microorganisms as chlorine, it does not dissipate into the air as quickly, which allows it more time to disinfect. Chloramine also generates fewer by-products than chlorine.
7. *Answers are as follows:*
   - trihalomethanes (THMs)—less than 0.08 mg/L
   - haloacetic acids (HAAs)—less than 0.06 mg/L
8. The main health hazard of excessive amounts of fluoride in the drinking water is dental fluorosis. Dental fluorosis is a condition in which white spots can form on a tooth surface for mild cases. Excessive enamel damage such as pitting may occur with excessively high consumption of fluoride.
9. *Answers are as follows:*
   - lead
   - arsenic
   - mercury
   - copper
10. Part 2: Water Data Record Keeping and Response of Feature 35, Periodic Water Quality Testing, requires that a policy be written concerning how to enforce water quality monitoring and record-keeping strategies. The policy must specify that detailed records of testing and inspections be kept for a minimum of three years. The policy must also include a detailed plan for taking action if the levels of dissolved metals exceed the limits set in Feature 31, Inorganic Contaminants.
11. Part 3: Microbial Elimination of Feature 36, Water Treatment, requires one of the two listed methods of microbial elimination: ultraviolet germicidal irradiation (UVGI) or filters rated by the National Science Foundation (NSF-rated filters) to remove microbial cysts.
12. *Answers are as follows:*
   - formation of Legionella management team
   - water system inventory and production of process flow diagrams
   - hazard analysis of water assets
   - identification of critical control points
   - maintenance and control measures, monitoring, establishment of performance limits, and corrective actions
   - documentation, verification, and validation procedures
13. Mild dehydration can result in muscle cramps, dry skin, and headaches. Severe dehydration can cause confusion, rapid heartbeat and breathing, shock, and delirium.
14. *Answers are as follows:*
   - aluminum—less than 0.2 mg/L
   - chloride—less than 250 mg/L
   - manganese—less than 0.05 mg/L
   - sodium—less than 270 mg/L
   - sulfate—less than 250 mg/L
   - iron—less than 0.3 mg/L
   - zinc—less than 5 mg/L
   - total dissolved solids—less than 500 mg/L
15. *Answers are as follows:*
   - daily cleaning of mouthpieces, guards, and basins to prevent lime and calcium buildup
   - quarterly cleaning of outlet screens and aerators to remove debris and sediment

WELL AP Exam Practice Questions __________ 99

1. D
2. A
3. B
4. A
5. D

CHAPTER 7—NOURISHMENT

Chapter Review ____________________ 120

Completion

1. macronutrient
2. macronutrients
3. Obesity
4. 25.0
5. sugar
6. pathogens
7. artificial ingredients
8. Food Advertising
9. 650
10. Organic
11. 20 L (0.7 ft^3)
12. 0.8 km (0.5 miles)
13. 25

Short Answer ________________________________ **120**

1. The six essential nutrients are proteins, carbohydrates, lipids, water, vitamins, and minerals. Nutrients are important to the aims of the Nourishment Concept because they are responsible for providing energy and helping the body grow, maintain, and repair its bones, muscles, organs, and cells.
2. *Answers are as follows:*
   - obesity
   - hypertension
   - cardiovascular disease

- diabetes
- cancers

3. *Answers are as follows:*
   - visible and accessible salad bar or salad-providing section
   - prominent visual displays of fruits and vegetables
   - vegetables placed at the front of the foodservice line
   - fruits placed at checkout locations
4. *Answers are as follows:*
   - peanuts
   - fish
   - shellfish
   - soy
   - milk and dairy products
   - egg
   - wheat
   - tree nuts
   - gluten
5. The column of water from the faucet to the bottom of the basin must be at least 25 cm (10″) in length and the basin must be at least 23 cm (9″) in width and length.
6. *Answers are as follows:*
   - total calories
   - macronutrient content, including total protein, fat, and carbohydrates
   - total sugar content
7. *Answers are as follows:*
   - Part 1: Cooking Material
   - Part 2: Cutting Surfaces
8. *Answers are as follows:*
   - circular plates with diameters no larger than 25 cm (10″)
   - noncircular plates with surface areas no larger than 507 cm^2 (79 in^2)
   - bowls no larger than 473 mL (16 oz)
   - cups no larger than 473 mL (16 oz)
9. *Answers are as follows:*
   - peanut-free
   - gluten-free
   - lactose-free
   - egg-free
   - no animal, seafood, or dairy products
   - no animal products except eggs and dairy
10. *Answers are as follows:*
    - planting medium
    - irrigation
    - lighting for indoor spaces
    - plants
    - gardening tools

WELL AP Exam Practice Questions 122

1. D
2. C
3. D
4. A
5. D

CHAPTER 8—LIGHT

Chapter Review

Completion 146

1. 400; 700
2. warmer
3. Flicker
4. Luminous flux
5. Luminance
6. circadian rhythm
7. teal-blue (≈480 nm)
8. 215 lux (20 fc)
9. equivalent melanopic lux (EML)
10. 53
11. electrochromic glass
12. Low-Glare Workstation Design
13. 80
14. Automated Shading and Dimming Controls
15. 75
16. 40

Short Answer 147

1. *Answers are as follows:*
   - ambient lighting
   - task lighting
   - accent lighting
2. The circadian rhythm can influence sleep/wake cycles, hormone release, body temperature, and metabolism, all of which regulate physiological processes such as digestion and sleep.
3. *Answers are as follows:*
   - rods: green-blue light (498 nm)
   - cones: green-yellow light (555 nm)
   - intrinsically photosensitive retinal ganglion cells (ipRGCs): teal-blue light (≈480 nm)
4. *Answers are as follows:*
   - maximum brightness contrasts between main rooms and ancillary spaces, such as corridors and stairwells, if present
   - maximum brightness contrasts between task surfaces and immediately adjacent surfaces, including adjacent visual display terminal screens
   - brightness contrasts between task surfaces and remote, nonadjacent surfaces in the same room
   - the way brightness is distributed across ceilings in a given room that maintains lighting variety but avoids both dark spots or excessively bright spots
5. $EML = L$ (lux) × R (melanopic ratio)
6. WELL differentiates between view and daylight windows by considering glazing below 2.1 m (7′) to be a view window and glazing above 2.1 m (7′) to be a daylight window.
7. *Answers are as follows:*
   - interior window shading or blinds
   - external shading systems
   - interior light shelves
   - a film of micromirrors on the windows that reflect light toward the ceiling
   - variable opacity glazing
8. *Answers are as follows:*
   - ceilings: LRV of 0.8 (80%) for 80% of surface area in regularly occupied spaces
   - vertical surfaces: LRV of 0.7 (70%) for 50% of surface area directly visible from regularly occupied spaces
   - furniture systems: LRV of 0.5 (50%) for 50% of surface area directly visible from regularly occupied spaces
9. Daylight modeling is the process of using a computer program to simulate the effects of natural and artificial light on the interior of a new or existing building. Daylight modeling allows project teams to compare various building design options to achieve the appropriate lighting levels for a given space. The minimum levels of natural lighting in a space are defined with spatial daylight autonomy (sDA), and the maximum levels of sunlight are defined with annual sunlight exposure (ASE).

WELL AP Exam Practice Questions 149

1. B
2. C
3. B
4. A
5. D

CHAPTER 9—FITNESS

Chapter Review

Completion 166

1. resistance
2. Sedentary behavior
3. two to four
4. 75
5. pedestrian amenity
6. 70
7. four
8. diverse use
9. 370 m^2 (4000 ft^2)
10. Active transportation
11. 5
12. cardiorespiratory
13. muscle-strengthening
14. 3
15. 60

Short Answer 167

1. Answers are as follows:
   - located within 7.5 m (25′) of the main entrance, main entry checkpoint, edge of the lobby, or main welcome area
   - clearly visible from the main entrance, main entry checkpoint, edge of the lobby, or main welcome area or must be located visually before any elevators upon entering from the main entrance
   - a minimum width of 1.4 m (56″), or the width allowable by local code, between handrails
2. *Answers are as follows:*
   - artwork
   - music
   - daylighting using windows or skylights at least 1 m^2 (10.8 ft^2) in size
   - view windows to the outdoors or building interior
   - light levels of at least 215 lux (20 fc) when the stairs are being used
   - biophilic elements
3. Feature 65, Activity Incentive Programs, is a precondition for New and Existing Interiors and New and Existing Buildings.
4. *Answers are as follows:*
   - Part 1: Fitness Programs
   - Part 2: Fitness Education
5. *Answers are as follows:*
   - bench
   - a cluster of movable chairs and tables
   - a drinking fountain or water refilling station
6. *Answers are as follows:*
   - water fountain or water feature
   - plaza or open-air courtyard
   - garden or other landscaped elements
   - public art
7. The external exercise spaces that can be used are a green space or park with playground features; a workout station or fitness zone; a trail network; an accessible body of water or public swimming pool; a gym, fitness, or training center; or a recreational field.
8. Bicycle maintenance tools and both short- and long-term bicycle storage must be available for occupant use.
9. Cardiorespiratory exercise equipment and muscle-strengthening exercise equipment are required by Feature 70, Fitness Equipment.
10. Active workstations are required for 3% or more of employees.

WELL AP Exam Practice Questions 169

1. A
2. C
3. C
4. D
5. B

CHAPTER 10—COMFORT

Chapter Review

Completion 190

1. Acoustics
2. musculoskeletal disorders
3. Olfactory
4. 55-2013
5. Accessible Design
6. 50
7. Noise criteria (NC)
8. RT60
9. Sound Masking
10. 0.9; 0.8
11. Doorway Specifications
12. Free address
13. 200 m^2 (2150 ft^2)
14. 10

Short Answer 191

1. *Answers are as follows:*
   - headaches
   - migraines
   - dizziness
   - allergic reactions
   - skin irritation
   - asthma attacks
   - mental distractions
   - increased stress
2. *Answers are as follows:*
   - metabolic rate
   - clothing insulation
   - air temperature
   - mean radiant temperature
   - air speed
   - humidity
3. *Answers are as follows:*
   - Part 1: Visual Ergonomics of Feature 73 requires that all computer screens, including laptops, be adjustable for height and distance from the user.
   - Part 2: Desk Height Flexibility of Feature 73 requires that at least 30% of workstations have the ability to alternate between sitting and standing positions using adjustable-height sit-stand desks, desktop height adjustment stands, or pairs of fixed-height desks of standing and seated heights.
   - Part 3: Seat Flexibility of Feature 73 requires that the workstation chair height adjustability be compliant with the HFES 100 standard or BIFMA G1 guidelines. Workstation seat depth adjustability must be compliant with the HFES 100 standard.
4. *Answers are as follows:*
   - open office spaces and lobbies that are regularly occupied and/or contain workstations—NC of 40
   - enclosed offices—NC of 35
   - conference rooms and breakout rooms—NC of 30 (NC of 25 recommended)

5. A mechanically ventilated building depends on a system of fans, heating units, and chillers to provide heating and cooling to occupied spaces. A naturally conditioned building depends on pressure differences to create airflow and provide heating and cooling to the spaces in the building.
6. *Answers are as follows:*
   - negative pressurization
   - interstitial rooms
   - vestibules
   - hallways
   - self-closing doors
7. Noise isolation class (NIC) is a field test for determining the sound transmitting abilities of a wall. Because NIC is a field test, it considers the full environment in which the wall is placed, not just the wall itself. This provides a more accurate measurement of sound cancellation compared to sound transmission class (STC) levels. Sound transmission class (STC) is a laboratory method for determining the sound transmission through a wall.
8. A hydronic radiant heating and/or cooling system is a radiant temperature system that uses water or another heat-transfer fluid to carry heated or chilled water from the point of generation to the point of use. An electric radiant system is a radiant temperature system that uses the heat created by the resistance of wiring to electrical current in order to warm the air or material around it.

WELL AP Exam Practice Questions ________ 193

1. B
2. A
3. C
4. C
5. A

CHAPTER 11—MIND AND INNOVATION

Chapter Review

Completion ________ 214

1. 20
2. charrette
3. Post-occupancy
4. Beauty and Design I
5. Biophilia
6. 186 m^2 (2000 ft^2)
7. 2700
8. 50
9. Workplace Family Support
10. 50
11. employee assistance program (EAP)
12. Organizational transparency
13. 929 m^2 (10,000 ft^2)
14. 25

Short Answer ________ 215

1. *Answers are as follows:*
   - a positive impact on personal and work relationships
   - higher productivity
   - better performance
   - less absenteeism
   - fewer workplace accidents
2. *Answers are as follows:*
   - metabolic syndrome
   - cardiovascular disease
   - gastrointestinal disorders
   - negative skin conditions
3. *Answers are as follows:*
   - decreased productivity
   - absenteeism
   - negative interactions between employees
4. *Answers are as follows:*
   - Perform a values assessment and alignment exercise within the team to inform any project goals as well as strategies to meet occupant expectations.
   - Discuss the needs of the occupants, focusing on wellness.
   - Set future meetings to stay focused on the project goals and to engage future stakeholders who join the process after the initial meeting, such as contractors and subcontractors.
5. *Answers are as follows:*
   - building site selection, including public transportation
   - the seven WELL Concepts
   - plans for implementing the project team's analyses and decisions
   - operations and maintenance plans for facility managers and wellness-related building policy requirements
6. *Answers are as follows:*
   - acoustics
   - thermal comfort (including humidity and airflow)
   - furnishings
   - workspace light levels and quality
   - odors, stuffiness, and other air quality concerns
   - cleanliness
   - layout
7. A project must include design elements that thoughtfully create unique and culturally rich spaces. These design elements must contain features intended for the following:
   - human delight
   - celebration of culture
   - celebration of spirit
   - celebration of place
   - meaningful integration of public art
8. *Answers are as follows:*
   - Part 1: Nature Incorporation
   - Part 2: Pattern Incorporation
   - Part 3: Nature Interaction
9. Adequate quality sleep is important for maintaining mental and physical performance. Poor sleep hygiene can lead to a higher risk of weight gain, depression, heart disease, high blood pressure, or stroke.
10. *Answers are as follows:*
    - body weight/mass
    - activity and steps
    - heart rate
    - sleep duration, quality, and regularity
11. *Answers are as follows:*
    - Declare® label
    - Health Product Declaration
    - Any method accepted in USGBC's MR Credit—Building Product Disclosure and Optimization—Material Ingredients, Option 1: Material Ingredient Reporting, from the LEED v4 BD+C rating system
12. Employees with companies that have high levels of organizational transparency and positive social equity practices are more satisfied, have a greater sense of loyalty, are more productive, and have a lower risk of poor health. The organization seeking WELL certification or compliance must meet one of the two following transparency programs:
    - JUST™ Program
    - G4 Sustainability Reporting Guidelines

13. *Answers are as follows:*
    - The ceiling height for a room with a width of 9 m (30′) or less must be at least 2.7 m (8.8′).
    - The ceiling height for a room with a width greater than 9 m (30′) must be at least 2.75 m (9′) plus 0.15 m (0.5′) for every additional 3 m (10′) of room width.
    - The ceiling height for a room with a full-wall window view to the outdoors or an atrium space must be at least 2.75 m (9′) if the room has a width of 12 m (40′) or less. If the room has a width over 12 m (40′), the ceiling height must have an additional 0.15 m (0.5′) for every additional 4.5 m (15′) of room width.
14. *Answers are as follows:*
    - Part 1: Outdoor Biophilia—the use of biophilic design elements in outdoor areas of the project site
    - Part 2: Indoor Biophilia—the use of biophilic design elements such as potted plants, planted beds, or plant walls in indoor spaces
    - Part 3: Water Feature—the use of biophilic water design in indoor spaces
15. *Answers are as follows:*
    - The proposed feature goes above and beyond the current requirements of an existing WELL feature.
    - The proposed feature relates to the concept of wellness in a novel way, is not already covered in the WELL Building Standard, and impacts the occupants or general public in a positive manner.
    - Innovations must be fully substantiated by scientific, medical, and industry research and explained in detail. These innovations must be consistent with all applicable laws, regulations, and leading building design and management practices.

WELL AP Exam Practice Questions ________ 218

1. A
2. D
3. A
4. B
5. A

CHAPTER 12—PRACTICE EXAM

Multiple Choice ________ 219

1. B
2. B
3. A
4. B
5. C
6. D
7. D
8. C
9. A
10. B
11. A
12. B
13. C
14. A
15. B
16. D
17. D
18. D
19. B
20. B
21. D
22. A
23. B
24. D
25. A
26. B
27. C
28. A
29. C
30. B
31. D
32. C
33. C
34. C
35. D
36. B
37. C
38. D
39. B
40. B
41. B
42. A
43. D
44. C
45. A
46. B
47. C
48. B
49. C
50. B
51. A
52. C
53. C
54. A
55. C
56. A
57. A
58. B
59. C
60. B
61. C
62. B
63. D
64. C
65. B
66. C
67. C
68. A
69. C
70. C
71. A
72. B
73. C
74. D
75. C
76. B
77. C
78. B
79. B
80. D
81. C
82. C
83. D
84. A
85. C
86. C
87. C
88. B
89. C
90. A
91. C
92. D
93. C
94. C
95. A
96. B
97. B
98. D
99. A
100. B

GLOSSARY

A

accent lighting: Lighting that is used to highlight architectural or design features of a space or to add visual interest to a space. Also known as decorative lighting.

acoustics: The study of sound and the properties of a space or building that determine how sound is transmitted or reflected within it or through it.

active transportation: A form of commuting by way of a physical activity such as biking or walking.

agricultural contaminant: A chemical pesticide, herbicide, or fertilizer that can be harmful to humans, animals, or the environment if it leaches into a water supply.

air exfiltration: The movement of air out of a conditioned building space.

air flushing: A technique used to remove or reduce airborne contaminants and pollutants by running the ventilation system for an extended period of time after construction is complete but before occupancy.

air infiltration: The movement of air into a conditioned building space through an unwanted void in the building envelope.

allergen: An environmental substance that can produce an allergic reaction in the body but may not be intrinsically harmful.

alternative adherence path (AAP): An alternative solution for meeting the intent of a WELL feature requirement.

altruism: The selfless act of helping another person.

ambient lighting: The main source of nonspecific illumination in a space. Also known as general lighting.

Americans with Disabilities Act (ADA): An enacted law that prohibits discrimination against persons with disabilities and ensures equal opportunity in employment, state and local government services, public accommodations, commercial facilities, and transportation.

annual sunlight exposure (ASE): A percentage of space in which the light level from direct sun alone exceeds a predefined threshold for some quantity of hours in a year.

arc tube: The light-producing element of an HID lamp.

appeal: A document that outlines a project team's disagreement with any finding of the WELL report or any decision regarding proposals for AAPs, curative actions, or the Innovation features.

asbestos: A naturally occurring mineral that was commonly used in insulation because of its chemical and flame resistance, tensile strength, and sound absorption properties.

atrazine: A pesticide that is among the most widely used pesticides in the United States to control broadleaf weeds in crops.

B

biomimicry: The imitation of natural or biological designs, patterns, and processes in the production of materials or structures.

biophilia: Human affinity for the natural world.

body mass index (BMI): A calculation based on height and weight to determine approximate body fat composition.

C

cancer: A group of diseases characterized by abnormal cell growth.

carbohydrate: Any of a group of organic compounds that includes sugars, starches, celluloses, and gums and serves as a major energy source to support bodily functions and physical activity.

carbon monoxide (CO): A colorless, odorless, and highly poisonous gas formed by incomplete combustion.

cardiorespiratory fitness: A component of physical fitness that involves the ability of the body, specifically the heart and lungs, to transport, absorb, and use oxygen during sustained physical activity. Also referred to as aerobic fitness.

cardiovascular disease: A class of medical conditions that affects the heart and blood vessels. Also known as heart disease.

charrette: An intensive, multiparty workshop that brings people together to explore, generate, and collaboratively produce building design options.

chloramine: A disinfectant formed when ammonia is added to chlorine and is commonly used as a secondary disinfectant in public water systems.

chlorine: A highly irritating, greenish-yellow gaseous halogen that can be introduced into a water supply as a gas, sodium hypochlorite solution, or calcium hypochlorite solid.

circadian rhythm: The internal clock that keeps the body's hormones and bodily processes on a roughly 24-hour cycle, even in continuous darkness.

coarse particle (PM_{10}): Particulate matter larger than 2.5 micrometers (μm) and smaller than 10 μm in diameter.

coliform: A microorganism that includes bacteria such as E. coli.

collaboration zone: A space within a building that encourages group interplay and discussion through its strategic layout and design.

color rendering index (CRI): An index that features a comparison of the appearance of 8 to 14 colors under a light source in comparison to a blackbody source of the same color temperature.

correlated color temperature (CCT): The spectral distribution of electromagnetic radiation of a blackbody at a given temperature, measured in degrees Kelvin (K).

curative action plan: A document that outlines strategies that a project team will employ to address unmet criteria as identified in the WELL report.

D

daylight window: Glazing that is specifically designed and located to allow for deeper penetration of daylight into the interior spaces of a building.

daylight modeling: The process of using a computer program to simulate the effects of natural light on the interior of a new or existing building.

decibel (dB): A unit of measurement for sound that is a logarithmic unit in which an increase in 10 decibels equals an increase by a factor of 10.

Declare label: A product declaration label designed by the International Living Future Institute that promotes transparency in disclosing material ingredients.

dedicated outdoor air system (DOAS): An HVAC system that provides 100% outside air directly to a building's zones for ventilation purposes.

diabetes: A group of diseases that impacts the metabolism due to insufficient insulin production (type 1) and/or high insulin resistance (type 2). Scientifically known as diabetes mellitus.

disinfectant: A chemical that is used to control or destroy harmful microorganisms as well as prevent their formation on inanimate objects and surfaces or in liquids.

disinfectant by-product (DBP): A compound that forms when chlorine and, to a slightly lesser extent, chloramine react with organic materials in a water supply.

displacement ventilation system: A ventilation system that introduces low-velocity supply air at a low level to displace warmer air that is then extracted at ceiling level.

diverse use: A distinct business or organization that provides goods or services intended to meet daily needs and is publicly available.

E

electric radiant system: A radiant temperature system that uses the heat created by the resistance of wiring to electrical current in order to warm the air or material around it.

electromagnetic radiation: Energy in the form of electromagnetic waves, including gamma rays, X-rays, ultraviolet (UV) light, visible light, infrared (IR) light, microwave radiation, and radio waves.

electromagnetic spectrum: The range of all types of electromagnetic radiation.

employee assistance program (EAP): An employer-sponsored workplace program that is designed to assist employees in identifying and resolving personal issues that may be affecting mental and emotional well-being, which may affect their job performance.

equivalent melanopic lux (EML): A measure of light that is used to quantify how much a light source will stimulate the light response of melanopsin.

ergonomics: The science of adapting objects, spaces, and processes, such as workstations and workflow, to accommodate people's capabilities in a safe and efficient manner.

F

feature: One of the 105 sections of the WELL Building Standard with a specific health intent.

fertilizer: A compound that contains nutrients that encourage the growth of a plant.

fine particle ($PM_{2.5}$): Particulate matter 2.5 μm or smaller in diameter.

fitness: The ability to carry out daily tasks with vigor and alertness, without undue fatigue, and with ample energy to enjoy leisure-time pursuits and respond to emergencies.

fluorescent lamp: A low-pressure discharge lamp in which ionization of mercury vapor transforms UV light generated by the discharge into visible light.

fluoride: A naturally occurring chemical that prevents or helps reverse tooth decay.

formaldehyde: A colorless gas compound that is used for manufacturing melamine and phenolic resins, fertilizers, dyes, and embalming fluids.

free address: The ability of occupants to choose their own work space within the office or workplace.

G

G4 Sustainability Reporting Guidelines: A reporting standard from the Global Reporting Initiative that helps organizations set goals, measure performance, and manage change in order to make their operations more sustainable for the environment, society, and the economy.

glare: The excessive brightness, excessive brightness contrast, and excessive quantity of light from a light source.

gluten: A type of protein found in the endosperm of cereal grains such as wheat, rye, barley, and spelt.

glyphosate: A nonselective herbicide that is used in many pesticide formulations, which may result in human exposure through its normal use due to spray drift, residues on food crops, and runoff into drinking water sources.

granular activated carbon (GAC) filtration system: A water filtration system that uses oxygen-treated carbon to chemically bond with the organic contaminants in water.

Green Business Certification Inc. (GBCI): A third-party organization that provides independent oversight of professional credentialing and project certification programs related to green building and health and wellness in the built environment.

H

haloacetic acid (HAA): A disinfectant by-product formed when chlorine or chloramine reacts with organic matter in water.

halogenated flame retardant: A chemical bonded with one of the halogen elements, such as chlorine or bromine, used in thermoplastics, thermosets, textiles, and coatings that inhibits or resists the spread of fire.

health literacy: The ability for people to obtain, study, and understand the health information and available health services that are communicated to them through health literature and presented options.

Health Product Declaration (HPD): An open-standard format from the Health Product Declaration Collaborative for reporting product content and associated health information for building products and materials.

herbicide: A type of pesticide that contains chemicals used to destroy or inhibit the growth of unwanted plants.

high-efficiency particulate air (HEPA) filter: A filter that removes 99.97% of all particles greater than 0.3 µm and satisfies standards of efficiency set by the Institute of Environmental Sciences and Technology (IEST).

high-intensity discharge (HID) lamp: A lamp that produces light from an arc tube.

high-touch surface: A surface that is frequently touched by building users and occupants.

humanely raised food: A meat, egg, or dairy product from an animal that has been kept, fed, and processed according to voluntary humane animal welfare standards.

hydronic heating and/or cooling system: A radiant temperature system that uses water or another heat-transfer fluid to carry heated or chilled water from the point of generation to the point of use.

hypertension: A condition characterized by high blood pressure.

I

illuminance: The amount of light incident on a given surface measured in lux or footcandles.

incandescent lamp: A lamp that produces light by the flow of current through a tungsten filament inside a sealed glass bulb sometimes filled with a gas.

inorganic contaminant: An element or compound that may be found in a water supply, occurring from natural sources such as the geology of a location, resulting from human activities such as mining and industry, or leaching into a water supply through outdated or malfunctioning water supply infrastructure.

integrative design process: A comprehensive approach to the design, construction, and operation of building systems and equipment that encourages a holistic approach to maximizing building performance, human wellness and comfort, and environmental benefits.

International Organization for Standardization (ISO): An independent, nongovernmental international organization that develops consensus-based, market-relevant standards for worldwide use.

International WELL Building Institute (IWBI): A public-benefit organization that administers the WELL Building Standard and aims to improve human health and well-being through the built environment.

isocyanate: An organic compound that is used in surface finishes and coatings.

J

JUST Program: A voluntary transparency labeling platform for organizations created by the International Living Future Institute that includes information on employee treatment, safety, social stewardship, and diversity.

K

kinetic degradation fluxion (KDF) filter: A water filter that contains flakes or granules of a copper-and-zinc alloy.

L

lamp: An output component that converts electrical energy into light.

lead: A naturally occurring metal found deep within the ground that was used in plumbing fixtures, lighting fixtures, and recycled building products.

Legionella: A bacterium that is found in freshwater and can cause a serious form of pneumonia called Legionnaires' disease.

light-emitting diode (LED) lamp: A lamp with semiconductor devices that produce visible light when an electrical current passes through them.

light reflectance value (LRV): A rating of how much usable light is reflected or absorbed by a given surface, with values ranging from 0 (0%) to 1 (100%).

lipid: An energy-providing nutrient made from fatty acids.

luminaire: A complete lighting unit that includes the components that distribute light, position and protect the lamps, and provide connection to the power supply.

luminance: A measurement of how bright a surface or light source will appear to the eye, measured in candela per square meter (cd/m^2) or footlamberts.

luminous flux: The total luminous output of a light source, measured in lumens and weighted to the visual sensitivity of the human eye.

luminous intensity: Radiant power weighted to human vision that describes light emitted by a source in a particular direction.

M

macronutrient: A nutrient needed by the body in large amounts.

malnutrition: A condition that results from insufficient nutrient intake, excess nutrient intake, or nutrient intake in the wrong proportions.

mechanical ventilation: Ventilation provided by mechanically powered equipment, such as motor-driven fans and blowers, but not by devices such as wind-driven turbine ventilators and mechanically operated windows.

mercury: A naturally occurring poisonous metal element that can be found in the earth's surface.

microbial cyst: A microorganism in its dormant state that is resistant to typical disinfection methods.

mineral: An inorganic nutrient that is required in small amounts to help regulate body processes.

minimum efficiency reporting value (MERV): A value assigned to an air filter that describes the amount of different types of particles removed when the filter is operating at the least effective point in its life.

musculoskeletal disorder: An injury or condition that affects the body's movements or its muscles, bones, tendons, nerves, or ligaments.

N

natural ventilation: The movement of air into and out of a space primarily through intentionally provided openings (such as windows and doors), through non-powered ventilators, or by infiltration.

nephelometric turbidity unit (NTU): The unit of measure for the turbidity of water.

noise criteria (NC): The sound pressure limits of the octave band spectra ranging from 63 Hz to 8000 Hz.

noise isolation class (NIC): A field test for determining the sound transmitting abilities of a wall.

noise reduction coefficient (NRC): The average value that determines the absorptive properties of materials.

nonpotable water: Water that is not fit for human consumption.

nutrient: A chemical that is required for metabolic processes, which must be taken from food or another external source.

O

obesity: A medical condition in which the accumulation of excess adipose tissue (body fat) poses an adverse effect on health.

off-gassing: The release of chemicals or particulates into the air from substances and solvents used in the manufacture of a building product.

olfaction: The sense of smell.

olfactory comfort: A person's perception of air quality within an environment based on their sense of smell.

optimization: An additional feature that can be used as a flexible pathway to achieve higher levels of WELL certification.

organic contaminant: A human-made compound or chemical containing carbon atoms that has leached into ground and surface water from industrial activities, such as the production of plastics.

organic food: Food produced without the use of chemically formulated fertilizers, growth stimulants, antibiotics, pesticides, or spoilage-inhibiting radiation.

organizational transparency: The full disclosure of information and open communication by an employer with other stakeholders, such as employees, and the public.

owner: An individual property owner, or a representative of the entity that owns the property, who is responsible for authorizing the registration of a project on WELL Online.

ozone (O_3): The triatomic form of oxygen that is hazardous to the respiratory system at ground level.

P

part: A requirement of a feature that dictates the parameters or metrics to be met.

particulate matter: A complex mixture of elemental and organic carbon, salts, mineral and metal dust, ammonia, and water that coagulate together into tiny solids and globules.

pathogen: An infectious biological agent such as a bacterium, virus, or fungus that is capable of causing disease in its host.

pedestrian amenity: A design feature that provides functional services for building occupants and visitors while at the same time making a space more comfortable and engaging.

perfluorinated compounds (PFCs): A family of fluorine-containing chemicals with unique properties to make materials stain- and stick-resistant.

performance verification: A site visit in which a WELL assessor conducts performance tests, visual inspections, and spot-checks, and that also includes follow-up analyses of collected data and samples from the site.

pesticide: A chemical that is used to destroy, repel, or control plants or animals.

photocatalytic oxidation system: An air sanitation system that uses a UV light along with a catalyst, usually titanium dioxide, to break down contaminants in the air stream.

phthalates: A group of chemicals used to make plastics more flexible and harder to break.

polybrominated diphenyl ethers (PBDEs): A group of brominated hydrocarbons that are used as flame retardants for plastics, foams, furniture and furnishings, textiles, and other household products.

polychlorinated biphenyl (PCB): A former commercially produced synthetic organic chemical compound that may be present in products and materials produced before the 1979 PCB ban.

polyurethane: A synthetic resin used chiefly in paints and varnishes.

potable water: Water that is fit for human consumption.

precondition: A feature that is mandatory for all levels of WELL certification.

processed food: Food that incurs a deliberate change before it is available for consumption.

process water: Water that is used for cooling towers, boilers, and industrial processes.

project administrator: The individual who acts as project manager and oversees the WELL certification process.

protein: An energy-providing nutrient that is made of carbon, hydrogen, oxygen, and nitrogen assembled in chains of amino acids.

psychrometric chart: A graphical representation of the properties of air at various conditions.

psychrometrics: The mathematical and graphical study of the properties of air.

R

radon: A radioactive, carcinogenic noble gas generated from the decay of natural deposits of uranium.

reverberation time: The time it takes for sound to decay, expressed in seconds.

reverse-osmosis (RO) filtration system: A water filtration system that uses a semipermeable membrane to filter water.

S

sedentary behavior: A manner of activity that involves sitting or lying down and is characterized by low levels of energy expenditure.

sick building syndrome (SBS): A set of symptoms, such as headaches, fatigue, eye irritation, and breathing difficulties, believed to be caused by indoor pollutants and poor environmental control that typically affects workers in modern airtight office buildings.

simazine: A popular herbicide that is used to control weeds.

sleep hygiene: A set of personal habits and practices that help maximize sleep quality.

sound: Energy composed of pressure waves, or vibrations, that travel through the air or another material and produce an audible signal when they reach a person's ear.

sound masking: The use of electronic devices to generate low-level background noise in order to mask sound distractions.

sound pressure level: The pressure variation associated with sound waves, usually measured in decibels (dB). Also known as acoustic pressure.

sound transmission class (STC): A laboratory method for determining the sound transmission through a wall.

spatial daylight autonomy (sDA): A percentage of floor space in which a minimum light level can be met completely for some proportion of regular operating hours by natural light.

spatial familiarity: A state of knowledge about an environment brought about by repeated association with elements within that environment.

stakeholder: Anyone who is affected, or will be affected, by the construction and operation of a building, such as the project owners, company shareholders, building occupants, visitors, construction and maintenance personnel, and even the surrounding community.

synergy: The interrelationship between systems or the components of those systems that can be realized through strategic integration to achieve high levels of building performance, human performance, and environmental benefits.

T

task lighting: Lighting that is used to illuminate an area in order to allow the performance of a specific function.

thermal comfort: The condition of satisfaction with the thermal environment.

trade-off: A factoring of strategies that makes one strategy achievable while another strategy becomes too challenging or cost-prohibitive.

trans fat: An unsaturated fatty acid with hydrogen atoms on opposite sides of a double carbon bond, which makes its structure similar to a saturated fat. Also known as partially hydrogenated oil or trans-fatty acid.

trihalomethane (THM): A disinfectant by-product that is formed when chlorine reacts with organic matter in water.

triple bottom line: The concept of sustainability that includes the financial, environmental, and social bottom lines of a project.

turbidity: The amount of cloudiness in a liquid caused by suspended solids that are usually invisible to the naked eye.

2,4-dichlorophenoxyacetic acid (2,4-D): A major herbicide that is likely to run off or leach into ground and surface water sources.

U

ultraviolet germicidal irradiation (UVGI): A sterilization method that uses UV light to break down microorganisms by destroying their DNA.

urea-formaldehyde: A low-cost thermosetting resin that is used in the wood product industry.

V

variable opacity glazing: Glazing that has the ability to control the amount of light passing through it. Also known as smart glass.

view window: Glazing that is designed to allow building occupants clear views of the outside.

visible transmittance (VT): The amount of light in the visible portion of the spectrum (roughly 400 nm to 700 nm) that passes through a glazing material. VT is measured as a percentage or number from 0 to 1.

visual acuity: The clarity or sharpness of vision.

vitamin: An organic nutrient that is required in small amounts to help regulate body processes.

volatile organic compound (VOC): A material containing carbon and hydrogen that evaporates and diffuses easily at ambient temperature and is emitted by a wide array of building materials, paints, wood preservatives, and other common consumer products.

W

Walk Score®: A measurement on a scale of 0 to 100 that takes into account a building inhabitant's physical output.

wavelength (λ): The distance between two points on a wave in which the wave repeats itself.

WELL Accredited Professional (WELL AP): An individual who possesses the knowledge and skill necessary to support the WELL certification process.

WELL assessor: An independent third party who conducts on-site performance tests, inspections, and spot checks, as well as reviews documentation in order to evaluate a project's eligibility for WELL certification.

WELL Building Standard: A performance-based system for measuring, certifying, and monitoring features of the built environment that impact human health and well-being through air, water, nourishment, light, fitness, comfort, and mind. Simply referred to as WELL.

WELL Online: The web-based portal for registering for the WELL AP exam and for completing the WELL certification process.

WELL report: A comprehensive report of a project that includes a feature-by-feature summary indicating whether the project team successfully provided documentation to verify the achievement of each feature or whether the project has successfully met the measurable criteria for specific features.

INDEX

Page numbers in italic refer to figures.

A

G

H

I

J

K

L

M

N

O

P